RMP UPDATE

2018

A Report of the Regional Monitoring Program for Water Quality in San Francisco Bay

NOTE TO READERS: The RMP produces *The Pulse of the Bay* in odd years, and the *RMP Update* in even years. In contrast to *The Pulse*, which focuses on Bay water quality and summarizes information from all sources, the *RMP Update* has a narrower and specific focus on highlights of RMP activities.

DIGITAL VERSIONS of all RMP Updates are available at: www.sfei.org/rmp/update

DIGITAL VERSIONS of all Pulses are available at: www.sfei.org/rmp/pulse

COMMENTS OR QUESTIONS regarding the RMP Update can be addressed to Dr. Jay Davis, RMP Lead Scientist, (510) 746-7368, jay@sfei.org.

SUGGESTED CITATION: San Francisco Estuary Institute (SFEI). 2018. RMP Update 2018. SFEI Contribution #906. San Francisco Estuary Institute, Richmond, CA.

To download this report please visit www.sfei.org/rmp/update

COVER: Inspecting a sediment grab sample. Photograph by Shira Bezalel.

PREFACE

The overarching goal of the Regional Monitoring Program for Water Quality in San Francisco Bay (RMP) is to answer the highest priority scientific questions faced by managers of Bay water quality.

The RMP is an innovative collaboration between the San Francisco Bay Regional Water Quality Control Board, the regulated discharger community, the San Francisco Estuary Institute, and many other scientists and interested parties.

The purpose of this document is to provide a concise overview of recent RMP activities and findings, and a look ahead to significant products anticipated in the next two years.

The report includes:

- a brief summary of some of the most noteworthy findings of this multifaceted Program;

- a description of the management context that guides the Program; and

- a summary of progress to date and future plans for addressing priority water quality topics.

CONTENTS

PROGRAM IMPACT

The **IMPACT** of the RMP on Management Decisions

Informing High Stakes Decisions

Billions of dollars are at stake in the decisions that society makes regarding activities that are directly intended to protect Bay water quality. The region has made huge investments to build and operate the infrastructure to collect and treat the region's sewage and industrial wastewater, and continued investment at a similar scale will be needed to maintain, upgrade, and operate this infrastructure to serve a growing Bay Area population. The region has spent and will continue to spend comparably large sums to manage stormwater and establish green infrastructure in our cities to capture stormwater and minimize its adverse water quality impacts on the Bay. Large investments have been and will be made to manage contaminated sediment in the Bay, both at sites identified for cleanup and for dredging to maintain channels for commercial and recreational vessels.

Billions more are riding on decisions regarding activities that influence Bay water quality as unintentional side-effects. Commercial product formulation and usage (including pesticides, pharmaceuticals, personal care products, electrical equipment, home furnishings, automobile components, and many, many others), sediment management, water supply management, energy production, and habitat restoration and management are all immense and essential enterprises that have a tremendous influence on Bay water quality.

More than money is at stake. Protecting the health of people that eat fish and shellfish from the Bay is one of the primary objectives of water quality managers. Cleanup plans for many contaminants are driven by this objective, as are decisions regarding advisories to promote safe consumption of fish from the Bay. Cleanup plans also aim to protect the health of fish, wildlife, and all of the aquatic species that live in the Bay.

The goal of the RMP is to collect data and communicate information about Bay water quality in support of all of these management decisions. The $3.6 million annual budget for the RMP is used judiciously so that these decisions on Bay water quality are informed by sound science.

▲ Near the Port of Oakland. Photograph by Shira Bezalel.

Regulatory Policies Informed by the RMP

Management of pollutant discharges to the Bay: wastewater, stormwater, dredged material

303(d) Listings

Total Maximum Daily Load Control Plans (TMDLs)

- San Francisco Bay Mercury TMDL
- Guadalupe River Mercury TMDL
- San Francisco Bay PCBs TMDL
- North Bay Selenium TMDL
- Suisun Marsh TMDL for Dissolved Oxygen and Mercury

Permits

- National Pollutant Discharge Elimination System (NPDES) wastewater discharge permit provisions
- Municipal Regional Stormwater Permit - Load reductions, green infrastructure planning
- Mercury and PCBs Watershed Permit for Municipal and Industrial Wastewater
- Nutrient Watershed Permit for Municipal Wastewater

Criteria

- Site-specific objectives and implementation plan for copper
- Nutrient numeric endpoint framework (under development)

Contaminant of Emerging Concern (CEC) Action Plans

Commercial product formulation and usage

- State legislative bans: microbeads, PBDEs, copper in brake pads
- State flammability standards for furniture and building materials: flame retardants
- State pesticide regulations: e.g., pyrethroids
- State Safer Consumer Products regulations
- Federal legislative bans: PCBs, microbeads
- Federal pesticide regulations: DDT, chlordane, dieldrin, diazinon, and chlorpyrifos
- County and local drug take-back ordinances and programs

Dredging and dredged material management

- Dredging and dredged material disposal permits through the Dredged Material Management Office
- Long-Term Management Strategy for the Placement of Dredged Material in the San Francisco Bay Region (LTMS)
- Essential Fish Habitat Agreement for Maintenance Dredging Conducted Under the LTMS Program
- Regional restoration plans

Public health protection

- Fish consumption advice and communication

Fishing pier at Paradise Beach County Park in Tiburon. ▶
Photograph by Shira Bezalel.

RMP Impact Summary:
Municipal Wastewater Dischargers

DECISIONS INFORMED BY THE RMP

- **Are treatment plant modifications or upgrades, or source reduction activities needed?**

  - **Which contaminants need to be reduced in municipal wastewater?**
    Examples of contaminants currently under consideration for reductions are nutrients, the pesticides fipronil and imidacloprid, and other contaminants of emerging concern.

  - **At which treatment plants are the reductions needed?**
    Different segments of the Bay vary greatly in their general characteristics, including in some cases their sensitivity to additional contaminant loads. The need for load reductions may therefore vary in different segments of the Bay.

  - **How much of a reduction is needed?**
    The goal of TMDLs and other control plans is to reduce concentrations in the Bay to levels that do not significantly impact beneficial uses. This requires a solid understanding of impairment and contaminant cycling in the Bay.

  - **What is the effect of the reductions or modifications on Bay water quality?**
    Monitoring is essential in demonstrating that load reduction efforts achieve the desired improvement in beneficial use attainment. Monitoring is needed to ensure that treatment plant modifications (e.g., implementation of reverse osmosis for water reuse) have no adverse impacts on beneficial uses.

- **Are actions needed for other pathways to reduce loads and impairment from contaminants found in municipal wastewater?** A holistic understanding of the relative importance of loads for all pathways is needed to optimize overall load reduction efforts.

REGULATIONS ADDRESSED

NPDES Permits

Mercury TMDL

PCBs TMDL

North Bay Selenium TMDL

Copper Site-Specific Objective (SSO) Implementation Plan

Nutrient Watershed Permit

Mercury and PCBs Watershed Permit

CEC Action Plans

Cyanide SSO Implementation Plan

Department of Toxic Substances Control (DTSC) Safer Consumer Product Regulations

Department of Pesticide Regulation (DPR) state pesticide regulations

USEPA Federal Insecticide, Fungicide, and Rodenticide Act

RMP Impact Summary:
Municipal Stormwater Dischargers

DECISIONS INFORMED BY THE RMP

- **Which contaminants need to be reduced in municipal stormwater?** Reductions of legacy contaminants are currently a primary focus of stormwater management attention, but other contaminants, including contaminants of emerging concern, may also need to be reduced.

- **How much load reduction effort is needed?** The goal of TMDLs and other control plans is to reduce concentrations in the Bay to levels that do not significantly impact beneficial uses. This requires a solid understanding of the linkage between stormwater and Bay impairment.

- **Which tributaries should be priorities for actions to reduce loads?** Different segments of the Bay encompass variable watershed source areas and related loads, and vary greatly in their general characteristics, including in some cases their sensitivity to additional contaminant loads. The need for load reductions may therefore vary for tributaries discharging to different segments of the Bay.

- **Which sources or source areas in watersheds should be targeted for load reductions?** Identifying the sources and source areas in watersheds to target is a major challenge in reducing stormwater loads.

- **What is the effect of load reductions or other stormwater management and watershed modifications on Bay water quality?** Monitoring and modeling are essential to demonstrating that load reduction efforts achieve the desired improvement in beneficial use attainment. Other activities in the watershed (e.g., land use changes or changes in chemical use) may also affect contaminant loads in either beneficial or adverse ways.

- **Are actions needed for other pathways to reduce loads and impairment from contaminants found in municipal stormwater?** A holistic understanding of the relative importance of loads for all pathways is needed to optimize overall load reduction efforts.

REGULATIONS ADDRESSED

NPDES Permits

Municipal Regional Stormwater Permit

Mercury TMDL

PCBs TMDL

North Bay Selenium TMDL

Copper Site-Specific Objective Implementation Plan

CEC Action Plans

DTSC Safer Consumer Product Regulations

DPR state pesticide regulations

USEPA Federal Insecticide, Fungicide, and Rodenticide Act

RMP Impact Summary:
Industrial Wastewater Dischargers

DECISIONS INFORMED BY THE RMP

- **Are treatment plant modifications or upgrades, or source reduction activities needed?**

  - **Which contaminants need to be reduced in industrial wastewater?** For example, the need for selenium reductions in refinery effluent was identified in the 1990s, and treatment upgrades implemented in the late 1990s achieved large reductions in selenium loads.

  - **At which treatment plants are the reductions needed?** Specific industrial discharges may contain higher levels of chemicals that may merit special attention. For example, sites where fire-fighting foams have been used may discharge higher levels of PFOS, a chemical of emerging concern present in older formulations. In addition, different parts of the Bay vary greatly in their general characteristics, including in some cases their sensitivity to additional contaminant loads. The need for load reductions may therefore vary in different segments of the Bay.

  - **How much of a reduction is needed?** The goal of TMDLs and other control plans is to reduce concentrations in the Bay to levels that do not significantly impact beneficial uses. This requires a solid understanding of impairment and contaminant cycling in the Bay.

  - **What is the effect of the reductions or modifications on Bay water quality?** Monitoring is essential in demonstrating that load reduction efforts achieve the desired improvement in beneficial use attainment. Monitoring is needed to ensure that treatment plant modifications (e.g., implementation of reverse osmosis for water reuse) have no adverse impacts on beneficial uses.

- **Are actions needed for other pathways to reduce loads and impairment from contaminants found in industrial wastewater?** A holistic understanding of the relative importance of loads for all pathways is needed to optimize overall load reduction efforts.

REGULATIONS ADDRESSED

NPDES Permits

Mercury TMDL

PCBs TMDL

North Bay Selenium TMDL

Copper SSO Implementation Plan

Mercury and PCBs Watershed Permit

CEC Action Plans

DTSC Safer Consumer Product Regulations

RMP Impact Summary:
Dredgers

DECISIONS INFORMED BY THE RMP

- **Where can contaminated dredged material be disposed?** RMP sediment data are the basis for the Dredged Material Testing Thresholds for mercury, polycyclic aromatic hydrocarbons (PAHs), and PCBs. These thresholds determine when bioaccumulation testing will be required for dredged material to be discharged at unconfined open water disposal sites in the Bay. RMP sediment data also serve as the basis for in-Bay dredged material disposal limits called for in the PCBs and mercury TMDLs.

- **Should dredged material be reused within the Bay and where?** Management of sediment as a resource in the Bay requires understanding of the volumes, types, locations, and environmental drivers of sediment input. The RMP performs extensive monitoring of suspended sediment concentrations along with monitoring of suspended sediment loads at select tributaries. The RMP also funds special studies to understand sediment transport within the Bay.

- **Should dredging practices be modified to prevent impacts to fish and benthic species?** The benthic communities of the Bay provide important foraging habitat for many fish species. The RMP performs studies to understand whether dredging practices have an impact on benthic species and habitats. The RMP also studies whether exposure to contaminants in dredged material poses a risk to fish.

REGULATIONS ADDRESSED

2011 Programmatic Essential Fish Habitat Agreement, Measure 1

2011 Programmatic Essential Fish Habitat Agreement, Measure 7

PCBs TMDL

Mercury TMDL

Long-Term Management Strategy

PROGRAM HIGHLIGHTS

The RMP **TOP 10:**
Recent Activities and Accomplishments

1 CECs: Non-targeted Analysis Heightens Interest in Stormwater

Investigations using non-targeted analysis to screen for contaminants of emerging concern (CECs) provide an inventory of compounds present in the Bay. Findings can be used to inform targeted chemical monitoring or toxicity studies. Non-targeted analysis is valuable for early identification of potentially problematic CECs, which can greatly facilitate effective management to minimize water quality impacts.

The RMP's most recent non-targeted analysis focused on identifying water-soluble compounds in Bay water and treated wastewater, and documented the presence of a number of unexpected contaminants. The most numerous and intense contaminant signals were associated with samples from a site in San Leandro Bay that is heavily influenced by urban stormwater, suggesting this pollution pathway merits further examination.

Contaminants identified at this stormwater-influenced site include a number of rarely-studied chemicals associated with vehicle tires and fluids, some of which are considered highly toxic to aquatic life. A major, multi-year effort to investigate these and other emerging contaminants in Bay Area stormwater begins this fall.

MORE INFORMATION:
RMP Fact Sheet: Non-Targeted Analysis of Water-Soluble Compounds Highlights Overlooked Contaminants and Pathways (2018)

On the 2017 water cruise. Photograph by Meg Sedlak. ▶

2 CECs: PBDEs Re-Classified as a Low Concern

The decline in polybrominated diphenyl ethers (PBDEs) in the Bay is a pollution prevention success story that highlights the power of policy decisions to protect the environment, and the power of monitoring to detect the improvement. RMP monitoring of PBDEs in the Bay was part of an extensive body of science that informed multiple regulatory and business decisions designed to ban or phase-out the use of these toxic flame retardants in California and the US. The RMP prioritizes Bay CECs using a tiered framework that guides management and monitoring. In 2017, a rigorous RMP review of current levels in the Bay led to reclassification of PBDEs from the "moderate concern" tier to the "low concern" tier. Status & Trends monitoring of key matrices will continue for at least two more cycles.

MORE INFORMATION:
RMP Planning Document: Contaminants of Emerging Concern in San Francisco Bay: A Strategy for Future Investigations. 2017 Revision. (2017)

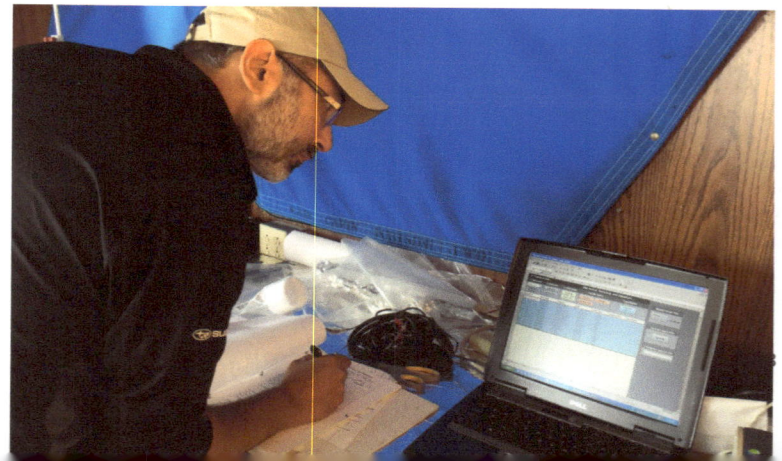

3 Small Tributaries: Guadalupe River Mercury Loads During High Flows of 2017

RMP monitoring is providing essential support for implementation of the TMDLs for mercury in San Francisco Bay and in the Guadalupe River. Efforts to reduce mercury loads to the Bay are primarily focusing on the Guadalupe River and urban stormwater. The Guadalupe River carries runoff from the New Almaden Mercury Mining District, historically the nation's largest mercury mining region and a continuing source of legacy contamination to the Lower South Bay. Load reduction activities in the Guadalupe watershed have been underway for over a decade and are planned to continue for at least another two decades. Guadalupe River flow has a major influence on mercury loading to the Bay, and the flow in the wet season of 2016/2017 was extremely high. RMP sampling during the high flows of January 2017 added to an extensive long-term dataset for loading from this watershed. The load measured during one rare once-in-five-years storm event was 70 kg, far more than the total wet season loads for every year except water year 2003 when a similar size storm last occurred. These estimates highlight the highly episodic nature of mercury transport from the watershed, which poses challenges for both monitoring and management.

MORE INFORMATION:
RMP Technical Report: Guadalupe River Mercury Concentrations and Loads During the Large Rare January 2017 Storm (2018)

4 PCBs: Assessment of Recovery from PCB Contamination in San Leandro Bay

The goal of RMP PCB special studies over the next few years is to inform the review and possible revision of the PCBs TMDL and the corresponding requirements in the reissued Municipal Regional Permit for Stormwater, both of which are tentatively scheduled to occur in 2020. Conceptual models are being developed for selected margin areas downstream of watersheds that are high priorities for management. The conceptual models will provide a foundation for establishing effective and efficient monitoring plans to track responses to load reductions, and will also help guide management actions.

A conceptual model for PCBs in San Leandro Bay was completed in 2018. Information from previous studies in this area was substantially augmented by an intensive field effort that evaluated the current status of contamination in sediment and fish. A simple mass budget model suggested that PCB concentrations in San Leandro Bay should respond to reductions in watershed loads. Sediment concentrations, however, have not declined since 1998, suggesting that continuing inputs are slowing recovery. Significant cleanup actions that have been recently completed or are happening soon on highly contaminated properties adjacent to San Leandro Bay should promote recovery. Watershed load reductions that focus on old industrial areas and small and moderate storms are predicted to have the greatest impact on recovery.

MORE INFORMATION:
RMP Technical Report: Conceptual Model to Support PCB Management and Monitoring in the San Leandro Bay Priority Margin Unit - Phase Three (2018)

RMP Technical Report: Gut Contents Analysis of Four Fish Species Collected in the San Leandro Bay RMP PCB Study in August 2016 (2018)

RMP Technical Report: San Leandro Bay Priority Margin Unit Study, Phase Two Data Report (2017)

5 Dioxins: Dioxin Synthesis

The Bay was placed on the California's 303(d) list of impaired waterways in 1998 as a result of elevated concentrations of dioxins and furans (commonly referred to as only 'dioxins') in fish. The RMP has monitored dioxins to support review of the 303(d) listing and establishment of a TMDL development plan or alternative. RMP dioxin studies from 2009-2014 addressed key information gaps in loadings from small tributaries, atmospheric deposition, concentrations in Bay sediment and water, and concentration trends in sport fish and bird eggs.

Sources of dioxin are expected to decrease nationally and locally due to past and current management efforts towards reducing emissions. Overall, it appears that dioxin will be a pollutant impairing beneficial uses in the Bay for a long time to come, with slow recovery, and modest progress on a Bay-wide scale unless interventions beyond the load reductions that are already underway are taken.

MORE INFORMATION:

RMP Technical Report: Current Knowledge and Data Needs for Dioxins in San Francisco Bay (2018)

6 Nutrients: Mussel Watch Monitoring and Expanded Advisory for Algal Toxins

Harmful algae can produce toxins that pose health risks to humans and wildlife. Monitoring for algal toxins is therefore critical for public health advisories. Since September 2015, SFEI has been conducting biweekly collections of naturally occurring mussels from 10 sites around the Bay, and collaborating with researchers at UC Santa Cruz to measure relevant phycotoxins (toxins produced by harmful algae). Multiple toxins have been regularly detected in mussels throughout Central and South Bay, including: domoic acid (amnesic shellfish poisoning toxin); saxitoxin (paralytic shellfish poisoning toxin); and microcystin (hepatotoxin). While domoic acid levels were always 100-200 fold lower than regulatory thresholds, saxitoxin and microcystin levels at some sites have exceeded regulatory thresholds. This monitoring led the California Department of Public Health to issue a consumption advisory for shellfish in Alameda and Contra Costa counties in March-May 2018. This advisory was in addition to the annual mussel quarantine normally in place from May 1 through October 31 anywhere on the California coast including bays, inlets, and harbors. The mussel-toxin data are also being used to inform mechanistic understanding of spatial and temporal variation of toxin sources.

MORE INFORMATION:

California Department of Public Health News Release: CDPH Warns Consumers Not to Eat Sport-Harvested Bivalve Shellfish from Alameda and Contra Costa Counties (March 2018)

Collaborator Journal Article: Blurred lines: Multiple freshwater and marine algal toxins at the land-sea interface of San Francisco Bay, California (2018)

◀ On the 2017 water cruise. Photograph by Meg Sedlak.

7 Nutrients: Dissolved Oxygen and Fish Habitat in Lower South Bay

Dissolved oxygen (DO) is a key water quality parameter that is often related to nutrient enrichment in estuaries around the world. In 2017-2018, SFEI launched a study to investigate the potential effects of DO on habitat quality in Lower South Bay (LSB) to inform decisions about nutrient regulation. The study approach involved convening a team of experts to advise on methods, analyzing high frequency DO measurements from seven mooring stations in LSB, and partnering with researchers from UC Davis to interpret several years of monthly fish abundance data in LSB relative to DO and other factors.

High-frequency DO measurements showed that low concentrations likely originate in sloughs and other margin habitats. In particular, sloughs that have been reconnected to former salt ponds appear to contribute to an especially high oxygen demand. The relationships between fish species abundances and DO were less clear. Species abundances exhibited substantial variability among the monthly trawls, and water quality variables also varied substantially. However, DO, temperature, and salinity strongly covary, making it difficult to make inferences about how fish abundances vary relative to any single water quality variable. To address these challenges, advanced statistical techniques (e.g., general additive models, GAMs) were used to disentangle the independent effects of DO, temperature, and salinity on fish abundances. The inital results suggest that the GAM approach could yield valuable information for some fish species. Proposed next steps include developing mechanistic models to predict DO concentrations throughout Lower South Bay and improving the statistical models of fish abundance.

MORE INFORMATION:
RMP Technical Report: Dissolved Oxygen in South San Francisco Bay: Variability, Important Processes, and Implications for Understanding Fish Habitat (2018)

8 Selenium: North Bay Selenium Studies

Studies under the RMP Selenium Strategy began in 2014, and have been aimed at establishing improved methods for obtaining information on impairment in the North Bay to support implementation of the North Bay Selenium TMDL.

Two multi-year sturgeon monitoring studies have been conducted. First, a non-lethal method of monitoring selenium concentrations in muscle biopsies from white sturgeon (the species and tissue established as the impairment indicator in the TMDL) was implemented in 2015-2017. These samples were collected in collaboration with a California Department of Fish and Wildlife tagging study. Selenium concentrations were significantly lower during the high flows of 2017 relative to the two prior drought years, confirming a pattern that was expected based on long-term trends in the clams that are a primary component of the sturgeon diet. Second, sturgeon were also monitored through coordination with an annual sturgeon fishing derby in the western Delta, also in 2015-2017. This study showed that selenium concentrations in muscle are correlated with concentrations in ovaries and liver (tissues that are more closely linked to fish health risk), and that concentrations in muscle plugs are well-correlated with concentrations in muscle fillets.

A third project developed a selenium monitoring design for the North Bay, with an emphasis on early detection of changes that could warrant changes in management approaches. Available data were reviewed and power analyses were conducted as a foundation for an integrated sampling design that includes monitoring of water, clams, and sturgeon.

MORE INFORMATION:
RMP Technical Report: North Bay Selenium Monitoring Design (2018)

RMP Technical Report: Selenium in Muscle Plugs of White Sturgeon from North San Francisco Bay, 2015-2017 (2018)

RMP Technical Report: Selenium in White Sturgeon from North San Francisco Bay: The 2015-2017 Sturgeon Derby Study (2018)

9 Sediment: RMP Goes All-In on Sediment Studies

Sediment delivered from watersheds is a precious and dwindling resource for the Bay. The goal for wetland restoration around the Bay is to restore 100,000 acres. In 2018, the San Francisco Bay Restoration Authority granted $18 million of Measure AA funds to work toward this goal. However, preliminary estimates show that there is not enough sediment to reach the restoration goal, even if all the sediment delivered to the Bay were used. The deficit gets even larger when sea level rise is considered. Therefore, knowing how much sediment is being delivered from which watersheds and changes in the supply over time is critical. In 2017, the U.S. Geological Survey teamed up with the RMP to answer this question. The ensuing report, published in 2018, shows that roughly 2 billion kilograms of sediment are typically delivered to the Bay, 63% of which comes from the small tributaries around the Bay versus 37% from the Delta and Central Valley. Bedload supply, after accounting for dredging, removals, storage in flood control channels, and errors in measurements, was indistinguishable from zero. Bedload is important because the heavier sediment that is transported along the bed is needed for beaches, wave breaks, and sand mining. This report listed dozens of recommendations for sediment monitoring and modeling that are needed to inform management decisions. Therefore, the RMP has formed a new workgroup to provide technical oversight and stakeholder guidance on RMP studies addressing questions about sediment delivery, sediment transport, dredging, and beneficial reuse of sediment.

MORE INFORMATION:
RMP Technical Report: Sediment Supply to San Francisco Bay, Water Years 1995 through 2016: Data, trends, and monitoring recommendations to support decisions about water quality, tidal wetlands, and resilience to sea level rise (2018)

10 Status and Trends: Monitoring the Bay Margins

The Bay "margins" are the mudflats and other very shallow areas around the edge of the Bay. The margins are key areas for contaminant impacts and monitoring, habitat restoration, dredged material re-use, and adaptation to sea level rise, so understanding these areas will be critical to protecting Bay water quality in the years to come. However, these areas have not historically been sampled during RMP cruises because they are too shallow for most research vessels. A RMP study conducted in 2015 and published in 2017 assessed the general level of contamination of PCBs and other contaminants in the margins of Central Bay, confirming the expectation that margin sediments have higher concentrations than the open Bay. In 2017, ambient concentrations in the large margin areas of South and Lower South Bay were monitored for PCBs, mercury, trace metals, and several emerging contaminants. A report on this study will be available by December 2018. After completion of this study, the RMP will have information on ambient concentrations in the margins everywhere south of the Richmond-San Rafael Bridge.

MORE INFORMATION:
RMP Technical Report: Characterization of Sediment Contamination in Central Bay Margin Areas (2017)

Collecting a water sample. Photograph by Shira Bezalel. ▶

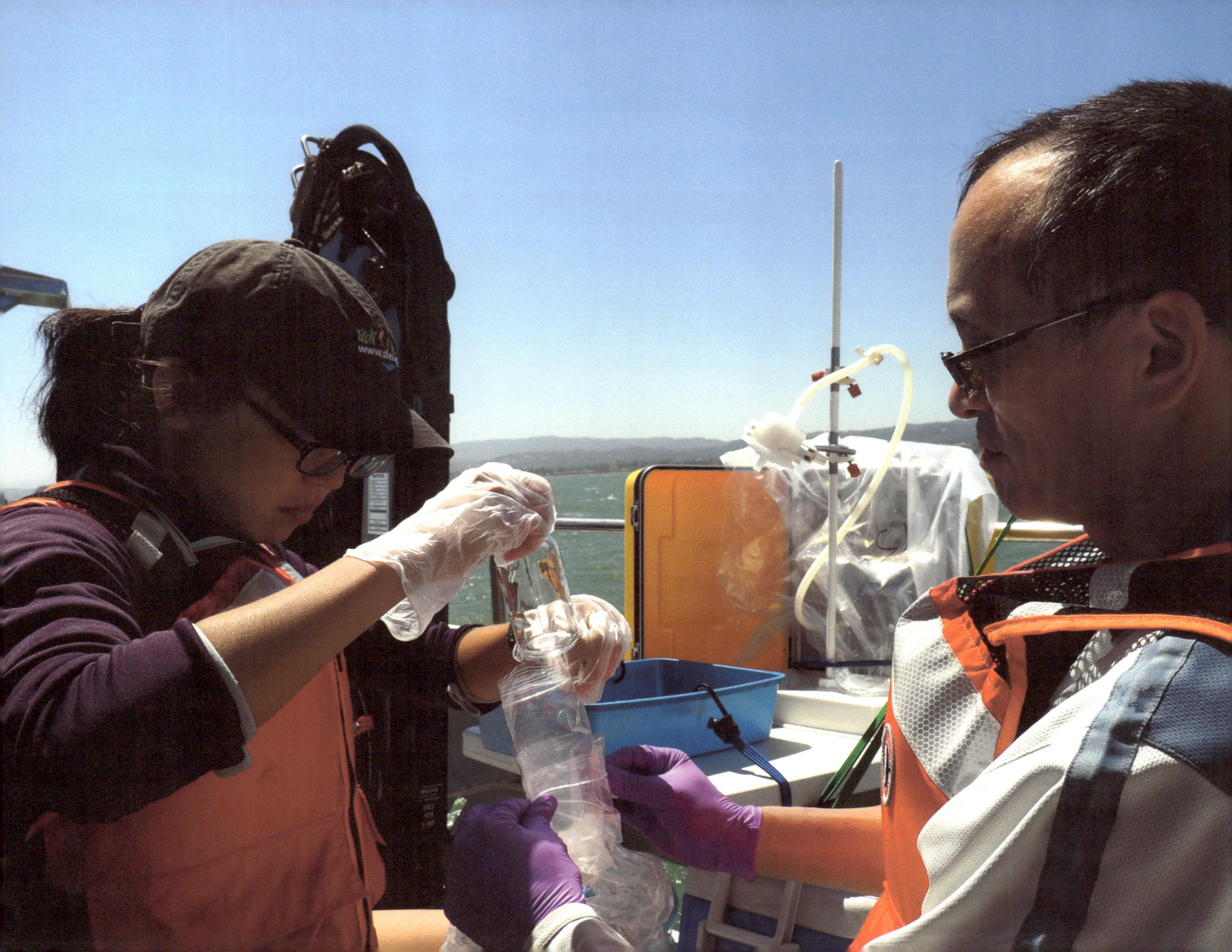

COMING **ATTRACTIONS**

1 **CECs: New Data and Risk Evaluations**

Water and sediment monitoring is providing information needed to evaluate the potential risks to Bay wildlife associated with plastic additives, alternative flame retardants, pesticides, and ingredients in fragrances added to personal care and cleaning products. Updates to the RMP Tiered Risk Framework for CECs will be highlighted in a 2019 CEC Strategy Update.

2 **CECs: Post-fire Non-targeted Sampling**

The RMP is supporting North Bay communities devastated by last fall's wildfires through an effort to scan post-fire stormwater runoff for CECs, complementing existing assessments for conventional contaminants. This non-targeted analysis will improve our understanding of the risks associated with toxic contaminants linked to wildfires. Results will be presented in 2019.

3 **Microplastic: Results and Policy Recommendations from Million Dollar Study**

A major Bay microplastic monitoring and modeling effort, funded primarily by the Gordon and Betty Moore Foundation with support from the RMP and others, is on its way to completion. We will distribute a report, educational materials, and policy recommendations via a public symposium in 2019.

4 **Small Tributaries: Advanced Data Analysis**

Reconnaissance data on PCBs in stormwater has provided good evidence to support management efforts in watersheds with high PCB concentrations. However, more insight can be obtained from the data for all watersheds through advanced data analysis. Enhanced ranking and fingerprinting methods for assessment of watersheds, subwatersheds, and PCB source areas will be applied to stormwater datasets to prioritize areas for management actions or further investigation.

5 **Small Tributaries: Remote Samplers for Pollutant of Concern Reconnaissance Monitoring**

Remote sediment samplers were pilot tested at 14 sites over the past four years and show promise as a lower-cost stormwater monitoring tool. Starting in 2019, these samplers will be used for initial characterization of new sites to determine if further storm sampling is warranted and to identify areas with high potential for cost-effective cleanup actions.

6 **Nutrients: Updated Science Plan and Funding Increase**

During its first five years, the Nutrient Management Strategy for San Francisco Bay has implemented major monitoring programs for dissolved oxygen, harmful algae, and nutrient concentrations, and developed biogeochemical models for the Bay and Delta. In 2019, funding for the program from the Bay Area Clean Water Agencies will more than double to $2.2 million per year. A Science Plan update in 2018 will chart the next 5-10 years of studies to answer priority management questions about nutrients in the Bay.

7 Sediment: Measuring Sediment Flux into Lower South Bay

In 2018, the US Geological Survey deployed new sensors to measure sediment fluxes at multiple depths and flocculation at the Dumbarton Bridge. The new measurements will provide managers with reliable information about how much sediment is being delivered to South Bay restoration projects. A basic report will be prepared in 2019, and a detailed interpretive report will be prepared in 2020.

8 Status and Trends: Sediment Monitoring 2018

The RMP monitors Bay sediment on a four-year cycle, with the latest round of sampling in the summer of 2018. Parameters to be measured will include PCBs, PAHs, PBDEs, fipronil, mercury and other metals. In addition, for efficiency, samples for special studies will be collected using the same research vessel.

9 Status and Trends: Sport Fish Monitoring 2019

The RMP monitors Bay sport fish on a five-year cycle, with the next round occurring in 2019. Parameters measured will include mercury, PCBs, selenium, dioxins, PFAS, and PBDEs. Non-targeted analysis of CECs will be performed, and samples will be archived for potential targeted analysis of microplastic and other CECs.

10 Pulse of the Bay 2019

The RMP publishes The Pulse of the Bay every other year, with the next edition coming in 2019.

Retrieving a sediment sample. Photograph by Meg Sedlak. ▶

PROGRAM OVERSIGHT

Collaboration and adaptation in the RMP are achieved through the engagement of stakeholders and scientists in frequent committee and workgroup meetings

The Steering Committee consists of representatives from discharger groups (wastewater, stormwater, dredging, industrial) and regulatory agencies (Regional Water Board and U.S. Army Corps of Engineers). The Steering Committee determines the overall budget and allocation of program funds, tracks progress, and provides direction to the Program from a manager's perspective.

Oversight of the technical content and quality of the RMP is provided by the **Technical Review Committee** (TRC), which provides recommendations to the Steering Committee.

Steering Committee

Technical Review Committee

San Francisco Bay Nutrient Management Strategy Committees

Nutrient Steering Committee

Microplastics Workgroup

Emerging Contaminants Workgroup

Exposure and Effects Workgroup

Sources, Pathways, and Loadings Workgroup

PCB/Dioxin Workgroup

Selenium Workgroup

Sediment Workgroup

Nutrient Technical Workgroup

Sport Fish Strategy Team

Small Tributary Strategy Team

PCB Strategy Team

Mercury Strategy Team*

*currently inactive

Workgroups report to the TRC and address the main technical subject areas covered by the RMP. The Nutrient Technical Workgroup was established as part of the committee structure of a separate effort — the Nutrient Management Strategy. This workgroup makes recommendations to the Nutrient Steering Committee on the use of RMP and other funds that support nutrient studies. The workgroups consist of regional scientists and regulators and invited scientists recognized as authorities in their field. The workgroups directly guide planning and implementation of special studies.

RMP Strategy Teams constitute one more layer of planning activity. These stakeholder groups meet as needed to develop long-term RMP study plans for addressing high priority topics.

PROGRAM MANAGEMENT

RMP FEES BY SECTOR: 2018

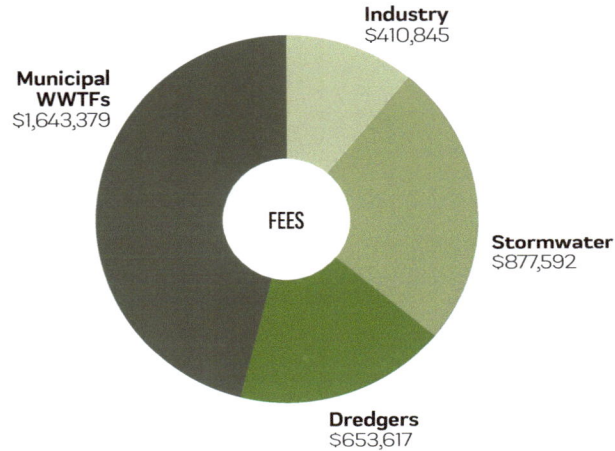

Industry $410,845

Municipal WWTFs $1,643,379

Stormwater $877,592

Dredgers $653,617

FEES

The fees target for 2018 was $3.59 million. 2018 was the first year without any fees from Cooling Water participants. The last of these participants, the Pittsburg Power Plant, ceased to have a discharge to the Bay in 2017.

COMMUNICATIONS

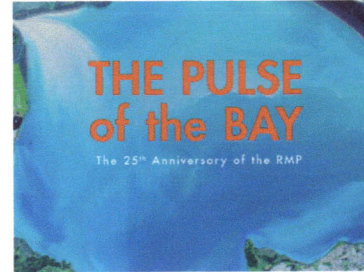

Includes the *Pulse of the Bay*, Annual Meeting, Multi-Year Plan, State of the Estuary report, RMP website, Annual Monitoring Report, technical reports, journal publications, newsletter, oral presentations, posters, and media outreach.

PROGRAM MANAGEMENT AND GOVERNANCE

Includes internal coordination (staff management), committee and workgroup meetings, coordination with Program participants, external coordination with related groups, program planning, contract and financial management, and workgroup and peer review coordination.

RMP EXPENSES: 2018

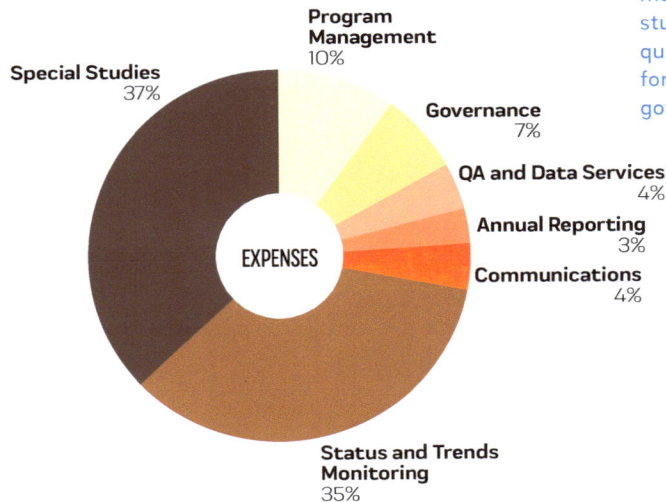

Program Management 10%

Special Studies 37%

Governance 7%

QA and Data Services 4%

Annual Reporting 3%

Communications 4%

Status and Trends Monitoring 35%

EXPENSES

RMP expenses in 2018 were 35% for status and trends monitoring, 32% for special studies, 11% for reporting and quality assurance, and 17% for program management and governance tasks.

DATA MANAGEMENT AND QUALITY ASSURANCE

The RMP database contains approximately 1.2 million records generated since the Program began in 1993. Web-based data access tools include user-defined queries, data download and printing functionality, maps of sampling locations, and visualization tools.

cd3 CONTAMINANT DATA DISPLAY & DOWNLOAD cd3.sfei.org

Per- and Polyfluoroalkyl Substances (PFAS) in San Francisco Bay Wildlife

AUTHORS: Sedlak M, Benskin J, Wong A, Grace R, Greig D. 2017.

TITLE: Per- and polyfluoroalkyl substances (PFASs) in San Francisco Bay wildlife: Temporal trends, exposure pathways, and notable presence of precursor compounds

JOURNAL: Chemosphere 185:1217-1226

Perfluoroalkyl and polyfluoroalkyl substances (PFAS) are a broad class of fluorine-rich specialty chemicals used for both industrial applications and consumer goods. Perfluoroalkyl substances are fully fluorinated, meaning that only fluorine atoms are bonded to the carbon backbone of the molecule. In contrast, polyfluoroalkyl substances are not fully fluorinated.

The RMP tiered prioritization framework breaks PFAS into subgroups with different levels of concern.

- **Perfluorooctane sulfonate (PFOS)** is classified as a **moderate concern** because it accumulates in the Bay food web to levels of potential concern for the health of wildlife and humans.

- **Perfluorooctanoic acid (PFOA)** and other long-chain perfluorocarboxylates were recently added to the **moderate concern** category because they also accumulate in Bay birds and seals to levels of potential concern and are not declining.

- **Other PFAS,** including short-chain perfluoroalkyl substances and polyfluoroalkyl substances, are classified as **possible concerns** due to detection in the Bay, potential for increased use, and a lack of information on toxicity to aquatic species. Some polyfluoroalkyl substances can degrade to perfluoroalkyl substances; these compounds are referred to as "precursors" of the perfluoroalkyl transformation products.

In 2017, Meg Sedlak and other RMP staff published a journal article summarizing RMP monitoring of PFAS in Bay wildlife. This information played a central role in the updated classification of PFAS in the tiered prioritization framework.

Harbor seals near Richardson Bay Marina. ▶
Photograph by Shira Bezalel.

Uses and Health Concerns

Over 4,700 PFAS are in use today in a wide variety of consumer, industrial, and commercial applications. PFAS have unique and highly desirable properties. Their ability to repel both oil and water means PFAS are frequently used as surface coatings for "water-proof" and "stain-proof" textiles, carpets, and food packaging materials. PFAS are also used as surface wetting agents in fire suppression foams, processing aids for the production of fluoropolymers like Teflon, and mist suppressants in metal-plating. The carbon-fluorine bonds in PFAS are some of the strongest known, which means many PFAS are extremely stable under all kinds of conditions. Unfortunately, this also means that they are extremely persistent once released to the environment.

In recent years, public concern has risen over the human health and ecological consequences of the widespread use of PFAS. Two of the most well-studied PFAS are perfluorooctane sulfonate (PFOS) and perfluorooctanoic acid (PFOA), eight carbon chain structures with a sulfonate and carboxylate at the end, respectively. Both PFOS and PFOA have been identified in drinking water supplies, sometimes substantially above human health thresholds. PFOS and PFOA have been associated with a myriad of adverse human health and ecological outcomes including suppression of the immune system, thyroid dysfunction, decreased birth weights, decreased sperm counts, testicular and kidney cancers, and ulcerative colitis, among others. Other PFAS have received much less scrutiny; however, structural similarities suggest that they may trigger similar concerns for their environmental fate and potential for adverse effects.

◄ Double-crested cormorants near Tiburon.
Photograph by Shira Bezalel.

Management of PFAS

In the US, production of PFOS was phased out by 2002, and production of PFOA and similar compounds was phased out by 2015. These federal actions were part of a broader international collaboration to reduce human and environmental risks associated with exposure to these compounds. For example, in 2009 many uses of PFOS were restricted under the Stockholm Convention on Persistent Organic Pollutants, a United Nations treaty.

Nonetheless, PFOS and PFOA production continues in other countries, such as China and India and there is concern that articles such as textiles and food packaging are treated with PFAS and then imported to the US. In addition, global production of related replacements, including short-chain perfluoroalkyl substances (containing four to six fluorinated carbons) and polyfluoroalkyl substances, means continuing use of and exposure to compounds that may potentially pose similar risks.

As the industry shifts to replacements such as the short-chain compounds, there is very little toxicological information about these alternatives available, and there is concern that they may be similarly problematic. While the short-chain perfluoroalkyl substances have much shorter half-lives in humans, they are more mobile in groundwater and less amenable to treatment via the sorption technologies that are typically employed to remove PFOS and PFOA from drinking water. Even less is known about the many members of the polyfluoroalkyl family, which have also seen increasing use as alternatives to PFOS and PFOA. With the exception of a handful of compounds, we do not know which specific polyfluoroalkyl substances are in use, making targeted analysis of environmental samples particularly challenging.

Because of the known risk associated with some of these chemicals, and the concern that many of the other members of this class may exhibit similar characteristics, several states are moving toward managing these compounds as a class. In 2018 the state of Washington passed the Healthy Food Packaging Act, which bans PFAS from food packaging materials. California is considering similar legislation. Under the California Safer Consumer Products Program, the Department of Toxic Substance Control (DTSC) has placed a high priority on evaluating the use of PFAS in carpets, rugs, and indoor upholstered furniture. DTSC considers this one of the largest potential sources of human exposure of PFAS from consumer products, and has proposed listing carpets and rugs containing PFAS as a Priority Product. Manufacturers of Priority Products are required to conduct an alternatives analysis; depending on the results, the chemical may be limited in its use.

In addition, some industries are concerned about the adverse environmental impacts from the use of these compounds, as well as consumer preference for PFAS-free product lines. Outdoor retailers such as Marmot, Fjällraven, Jack Wolfskin, Columbia, and Vaude now offer rain apparel that has not been treated with PFAS. IKEA has removed PFAS-treated textiles from all of its product lines, and Tarkett, a manufacturer of flooring, has removed PFAS from carpets. But in general the progress is slow.

In addition, even if these chemicals are successfully removed from the marketplace, it is likely that environmental exposure will continue, as there are significant reservoirs of residual PFOS, PFOA, and related compounds in both the environment and in urban settings as a result of decades of use.

A Wealth of RMP Data

PFAS are detected in the environment worldwide. Concerned that these compounds might be present in the Bay, the RMP began by monitoring PFAS in cormorant eggs in 2006 and archived Pacific harbor seal blood samples from 2004. PFOS was the primary PFAS identified, with concentrations of other PFAS an order of magnitude lower. Concentrations of PFOS in eggs and harbor seal blood from the South Bay were some of the highest observed relative to birds and seals from other monitoring sites around the world (Sedlak and Grieg 2012).

Over the last decade, the RMP has continued to track concentrations of PFOS and a handful of other PFAS in sport fish, bird eggs, and seals. In general, the short-chain alternatives such as PFBA are not detected in biota, suggesting that these compounds do not bioaccumulate.

Sport Fish

The RMP has monitored sport fish twice for 13 PFAS at five popular recreational fishing locations around the Bay. The 2009 sampling indicated low concentrations of PFOS and PFOSA, a precursor to PFOS, in sport fish; however, analytical methods improved substantially by the next sampling event in 2014, and PFOS was detected in approximately 75% of the fish, ranging in concentrations from 1.9 to 7.0 ng/g wet weight (ww) (Figure 1).

The 2014 sampling included a small pilot study in Artesian Slough in the Lower South Bay. The five sport fish caught from this area contained PFOS ranging in concentration from 2.5 to 17.2 ng/g. In addition, for the first time, there were detections of a number of the long-chain carboxylate compounds. This may reflect closer proximity to potential pollution pathways such as wastewater effluent and stormwater runoff.

California has not established a human health advisory tissue level for PFAS in fish. Michigan and Minnesota have established guidelines for PFOS based on the frequency of consumption. All sport fish analyzed were below the Minnesota threshold for unlimited fish consumption of 40 ng/g ww; 30% of the fish from the South Bay exceeded a sport fish consumption guideline of 9 ng/g ww for people who consume more than 4 meals per week. None of the sport fish from the Central Bay exceeded any threshold.

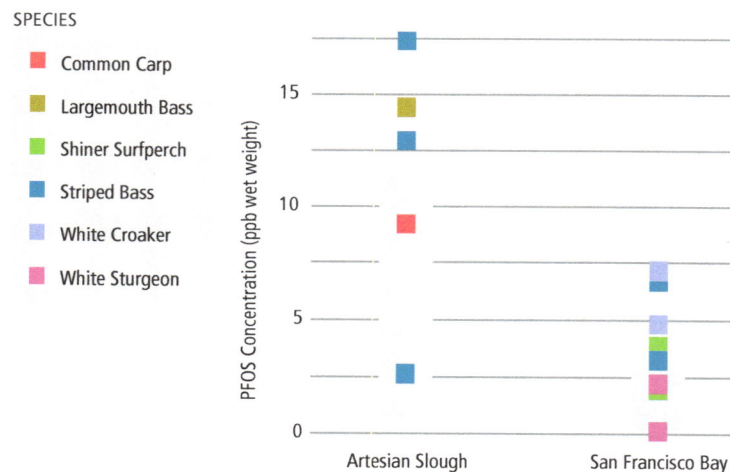

Figure 1. Concentrations of PFOS in Sport Fish. Points represent an individual fish for carp or bass from Artesian Slough and composite samples (all other species, including striped bass from San Pablo Bay and Central Bay) samples. Fish that did not contain PFOS above detection limits are not included. The San Francisco Bay fish were collected in 2014; Artesian Slough fish were collected in 2015. Xx use different symbols for individuals and composites

Cormorant Eggs

Since 2006, the RMP has regularly monitored Double-crested Cormorant *(Phalacrocorax auritus)* eggs for 13 PFAS from three open water site in the Bay.

PFOS has been the dominant PFAS identified (Figure 2); the remaining PFAS are an order of magnitude lower. PFOS concentrations were highest in South Bay, followed by Richmond Bridge (Central Bay), and then Wheeler Island (North Bay). Cormorant eggs from the South Bay have had some of the highest concentrations of PFOS observed globally, although over time, they appear to be declining. Given the relatively short half-life of PFOS in birds of two to three weeks, these results suggest on-going exposures. In contrast to PFOS, the concentrations of the long-chain perfluoroalkyl carboxylates like PFOA do not show discernible declining trends.

In the past, PFOS concentrations in South Bay eggs have exceeded a predicted no effects concentration (PNEC) of 1,000 ng/ml. However, more recent studies suggest that effects can occur at much lower levels. Field studies have indicated an approximately 50% reduction in hatching success of tree swallows at an egg PFOS concentration of 500 ng/g ww; a 15% reduction in success was observed at concentrations as low as 148 ng/g. The sensitivity of avian species to chemicals can vary widely, but the swallow studies suggest that PFOS concentrations in South Bay cormorant eggs are a potential concern. Information on the toxicity of other PFAS is virtually nonexistent.

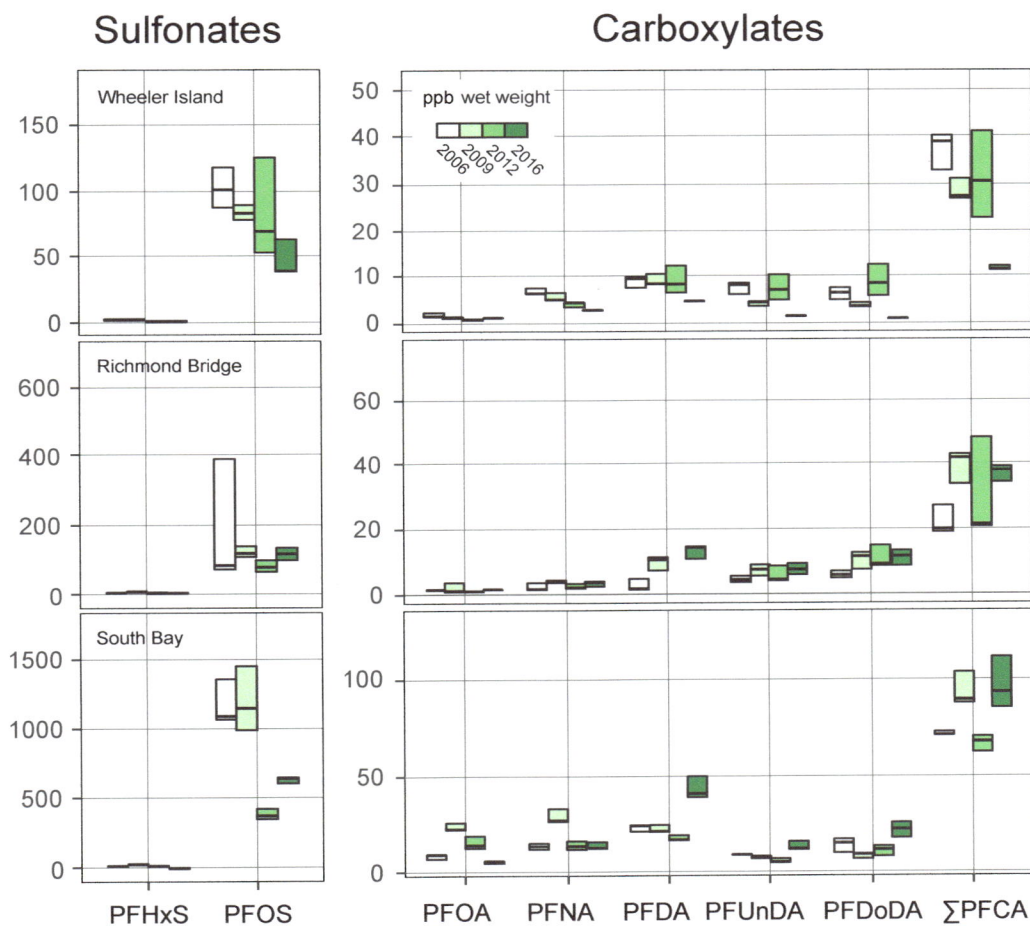

Figure 2. Concentrations of PFAS (ppb wet weight) in San Francisco Bay Double-crested Cormorant Eggs. Concentrations of perfluoroalkyl sulfonates (left panels) and perfluoroalkyl carboxylates (right panels) in Cormorant Eggs collected in 2006 (pale green), 2009 (light green), 2012 (green), and 2016 (dark green) from Wheeler Island, Richmond Bridge, and South Bay. Concentrations are plotted as boxplots, with the box denoting the 25th and 75th percentiles. The horizontal line within the box is the median concentration.

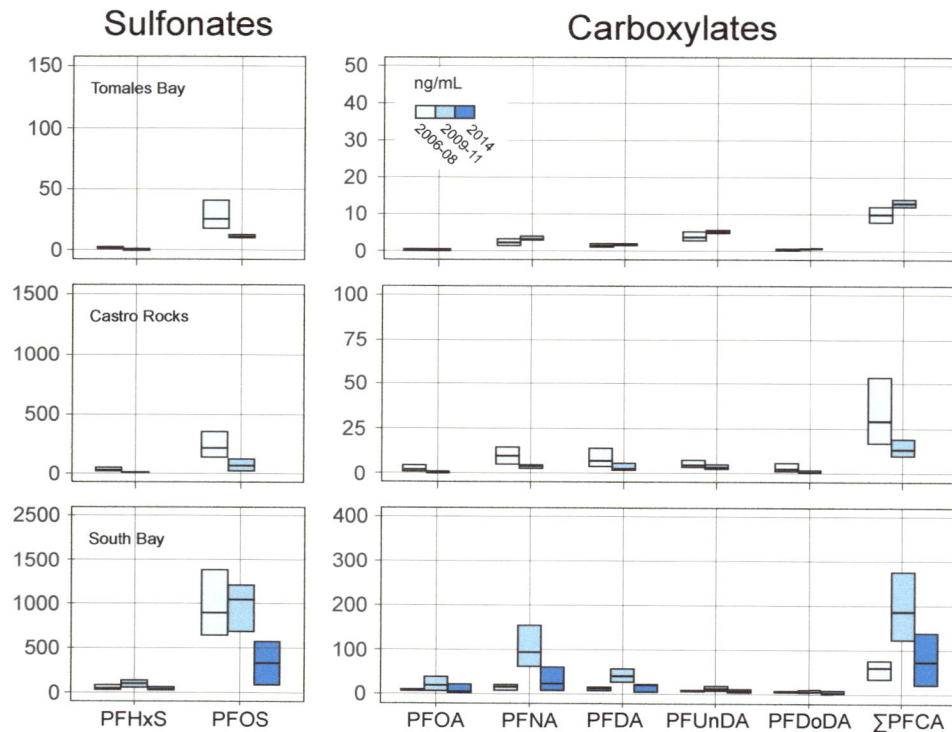

Figure 3. Concentrations of PFAS (ng/ml) in San Francisco Bay Seal Blood. Concentrations of perfluoroalkyl sulfonates (left panels) and perfluoroalkyl carboxylates (right panels) in seal blood collected between 2004-2008 (light blue), 2009-2011 (blue), and 2014 (dark blue) from Tomales Bay, Castro Rocks, and South Bay. In 2014, samples were only collected from South Bay. Concentrations are plotted as boxplots, with the box denoting the 25th and 75th percentiles. The horizontal line within the box is the median concentration.

Harbor Seals

The Pacific harbor seal *(Phoca vitulina richardii)* is a year-round resident of San Francisco Bay and surrounding coastal waters. Seals are primarily observed in the central and southern portions of the Bay, where they forage and haul-out on rocky shoals or mudflats to rest, birth, and nurture young pups. Seals have surprisingly high site fidelity, making them a highly desirable apex predator to monitor for PFAS.

The RMP has analyzed PFAS in blood from Bay harbor seals from 2004 through 2014 at Castro Rocks (beneath the Bay Bridge), in the Lower South Bay (Mowry and Corkscrew Slough), and at a reference site in Tomales Bay. The RMP has targeted blood because it has been shown that PFAS preferentially accumulate in blood relative to other tissues. In addition, sampling blood enables the RMP to track temporal trends in a non-invasive manner.

PFOS is the dominant PFAS detected in seals (Figure 3). The highest concentrations are observed in the South Bay. In fact, the initial studies of South Bay seals (2004 to 2008) had some of the highest concentrations observed world-wide, with a geometric mean of 906 ng/g ww and a range of 401 to 1,960 ng/g.

The remaining long-chain perfluoroalkyl carboxylates are typically an order of magnitude lower than PFOS concentrations. Similar to bird eggs, concentrations of PFOS have declined in the South Bay; however, concentrations of the long-chain perfluoroalkyl carboxylates show no discernible trends.

Very few studies have evaluated the toxicological effects of PFAS on seals. In a study of wild seals *(Pusa sibirica)* from Lake Baikal, Russia, concentrations on the lower end of those observed in San Francisco Bay seals were found to adversely affect gene function and immune response. Suppression of immune response has been widely observed in other mammals including rats, sea otters, and humans.

Inference from other mammals to seals is challenging given the current state of knowledge; however, in general, there appears to be some similarity in adverse effects observed among different mammals. Therefore, in the absence of established toxicity thresholds for harbor seals, it may be informative to evaluate these concentrations in light of human health thresholds. A recent review of human epidemiological studies of PFOS and PFOA and laboratory studies of mice concluded that blood serum levels of 1.3 ng/mL for PFOS and 0.8 ng/mL for PFOA elicit responses in humans relative to background concentrations. The levels observed in South Bay seals are well above these "benchmark dose levels."

Next Steps

The RMP has classified PFOS, PFOA, and long-chain carboxylates as a moderate concern based on the RMP's tiered prioritization framework that guides monitoring and management actions on emerging contaminants in the Bay. As such, a high priority is placed on monitoring these chemicals. The remaining PFAS have been classified as possible concern under the tiered prioritization framework due to uncertainty regarding their toxicity.

The RMP Status and Trends program will continue to monitor bird eggs and sport fish for 13 PFAS, including PFOS and PFOA, to track temporal trends and to assess the efficacy of the management actions to phase-out PFOS and PFOA. In addition, a 2019 study evaluating PFAS and a variety of other CECs in stormwater is planned. This information will help to assess shifts in the marketplace to short-chain PFAS and other alternatives. In the future, the RMP may use novel non-targeted analytical methods to qualitatively identify PFAS in the Bay.

RMP data will inform policies on the management of PFAS including the DTSC Safer Consumer Product program evaluation of PFAS in carpets and rugs; OEHHA's proposition 65 listing of PFOS and PFOA; and efforts by regulators and non-governmental agencies to manage these compounds as a class.

Additional Information

Sedlak, M. and D. Grieg. Perfluoroalkyl compounds (PFCs) in wildlife from an urban estuary. 2012. Journal of Environmental Monitoring 14: 146-154.

Sedlak, M., Sutton R., Wong A., Lin, Diana. 2018. Per and Polyfluoroalkyl Substances (PFAS) in San Francisco Bay: Synthesis and Strategy. SFEI Contribution No. 867. San Francisco Estuary Institute, Richmond CA.

Lin, D., Sutton, R., Shimabuku, I., Sedlak, M., Wu, J., and Holleman, R. 2018. Contaminants of Emerging Concern in San Francisco Bay: A Strategy for Future Investigations - 2018 Update. SFEI Contribution No. 873. San Francisco Estuary Institute, Richmond CA.

Recent **Publications**

The Pulse

The Pulse of the Bay: The 25th Anniversary of the RMP. SFEI. 2017. https://www.sfei.org/rmp/pulse

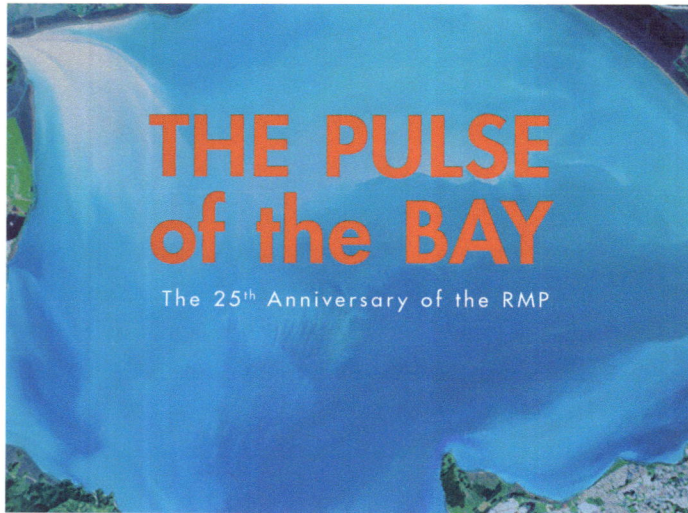

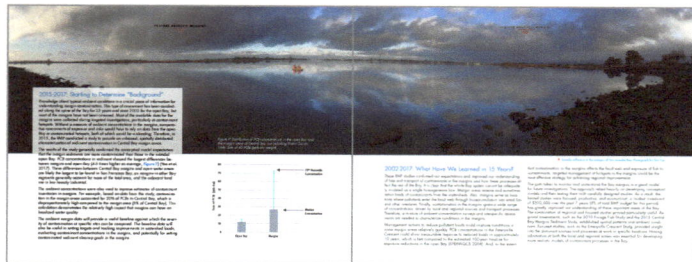

Journal Publications

Blurred lines: Multiple freshwater and marine algal toxins at the land-sea interface of San Francisco Bay, California. Peacock, M.B., Gibble, C.M., Senn, D.B., Cloern, J.E., and Kudela, R.M. 2018. Harmful Algae, 73: 138-147. https://www.sfei.org/documents/blurred-lines-multiple-freshwater-and-marine-algal-toxins-land-sea-interface-san-francisco

From Sediment to Top Predators: Broad Exposure of Polyhalogenated Carbazoles in San Francisco Bay (U.S.A.). Wu, Y., Tan, H., Sutton, R., and Chen, D. 2017. Environmental Science and Technology, 51: 2038-2046. http://www.sfei.org/documents/sediment-top-predators-broad-exposure-polyhalogenated-carbazoles-san-francisco-bay-usa

Long-term variation in concentrations and mass loads in a semi-arid watershed influenced by historic mercury mining and urban pollutant sources. McKee, L.J., Bonnema, A., David, N., Davis, J.A., Franz, A., Grace, R., Greenfield, B.K., Gilbreath, A.N., Grosso, C., Heim, W., et al. 2017. Science of The Total Environment, 605-606: 482-497. https://www.sfei.org/documents/long-term-variation-concentrations-and-mass-loads-semi-arid-watershed-influenced-historic

Microplastic pollution is widely detected in US municipal wastewater treatment plant effluent. Mason, S.A., Garneau, D., Sutton, R., Chu, Y., Ehmann, K., Barnes, J., Papazissimos, D., and Rogers, D.L. 2016. Environmental Pollution, 218: 1045-1054. http://www.sfei.org/documents/microplastic-pollution-widely-detected-us-municipal-wastewater-treatment-plant-effluent

Passage of fiproles and imidacloprid from urban pest control uses through wastewater treatment plants in northern California. Sadaria, A.M., Sutton, R., Moran, K.D., Teerlink, J., Brown, J.V., and Halden, R.U. 2016. Environmental Toxicology and Chemistry, 36: 1473-1482. http://www.sfei.org/documents/passage-fiproles-and-imidacloprid-urban-pest-control-uses-through-wastewater-treatment

Per- and polyfluoroalkyl substances (PFASs) in San Francisco Bay wildlife: Temporal trends, exposure pathways, and notable presence of precursor compounds. Sedlak, M.D., Benskin, J. P., Wong, A., Grace, R., and Greig, D.J. 2017. Chemosphere, 185: 1217-1226. http://www.sfei.org/documents/and-polyfluoroalkyl-substances-pfass-san-francisco-bay-wildlife-temporal-trends-exposure

Program Planning Documents

2018 Quality Assurance Program Plan for the Regional Monitoring Program for Water Quality in San Francisco Bay. Yee, D., Franz, A., Wong, A., and Trowbridge, P. 2018. https://www.sfei.org/documents/2018-quality-assurance-program-plan-regional-monitoring-program-water-quality-san

Charter: Regional Monitoring Program for Water Quality in San Francisco Bay. Trowbridge, P. 2017. http://www.sfei.org/documents/charter-regional-monitoring-program-water-quality-san-francisco-bay-0

Contaminants of Emerging Concern in San Francisco Bay: A Strategy for Future Investigations - 2018 Update. Lin, D., Sutton, R., Shimabuku, I., Sedlak, M., Wu, J., and Holleman, R. 2018. https://www.sfei.org/documents/contaminants-emerging-concern-san-francisco-bay-strategy-future-investigations-2018-update

Microplastic Monitoring and Science Strategy for San Francisco Bay. Sutton, R., and Sedlak M. 2017. http://www.sfei.org/documents/microplastic-monitoring-and-science-strategy-san-francisco-bay

RMP Small Tributaries Loading Strategy: Trends Strategy 2018. Wu, J., Trowbridge, P., Yee, D., McKee, L., and Gilbreath, A. 2018. https://www.sfei.org/documents/rmp-small-tributaries-loading-strategy-trends-strategy-2018

Summary of Workshop on Monitoring for Acidification Threats in West Coast Estuaries: A San Francisco Bay Case Study. Trowbridge, P., Shimabuku, I., Wheeler, S., Knight, E., Nielsen, K., Largier, J., Sutula, M., Valiela, L., and Nutters, H. 2017. http://www.sfei.org/documents/summary-workshop-monitoring-acidification-threats-west-coast-estuaries-san-francisco-bay

Fact Sheets, Selected Posters, and Presentations

Identifying and Addressing Contaminant Sources Impacting an Urban Estuary. Sutton, R., Sedlak, M., and Houtz, E. 2016. Poster. http://www.sfei.org/documents/identifying-and-addressing-contaminant-sources-impacting-urban-estuary

Optimizing Sampling Methods for Monitoring Pollutant Trends in San Francisco Bay Urban Stormwater. Melwani, A., Yee, D., Gilbreath, A., Davis, J., and McKee, L. 2016. Poster w https://www.sfei.org/documents/optimizing-sampling-methods-monitoring-pollutant-trends-san-francisco-bay-urban-stormwater

Pollutants of Concern Monitoring: A low-intensity, budget conscious stormwater sampling method to identify highly polluted areas for potential management action. Gilbreath, A., McKee, L., Hunt, J., and Yee, D., 2017. Presentation. https://www.sfei.org/documents/pollutants-concern-monitoring-low-intensity-budget-conscious-stormwater-sampling-method

Simple Mass Budget Model to Evaluate Long Term PCB Fate in the Emeryville Crescent Sub-embayment. Yee, D., Davis, J.A., Gilbreath, A.N., and McKee, L.J. 2016. Poster. http://www.sfei.org/documents/simple-mass-budget-model-evaluate-long-term-pcb-fate-emeryville-crescent-sub-embayment

Find links to all presentations from the 2016 RMP Annual Meeting on the 2016 RMP Annual Meeting webpage.

Find links to all presentations from the 2017 RMP Annual Meeting on the 2017 RMP Annual Meeting webpage.

Find links to all presentations from the 2018 RMP Annual Meeting on the 2018 Annual Meeting webpage.

Acoustic-release device and caged bivalves. Photograph by Paul Salop. ▶

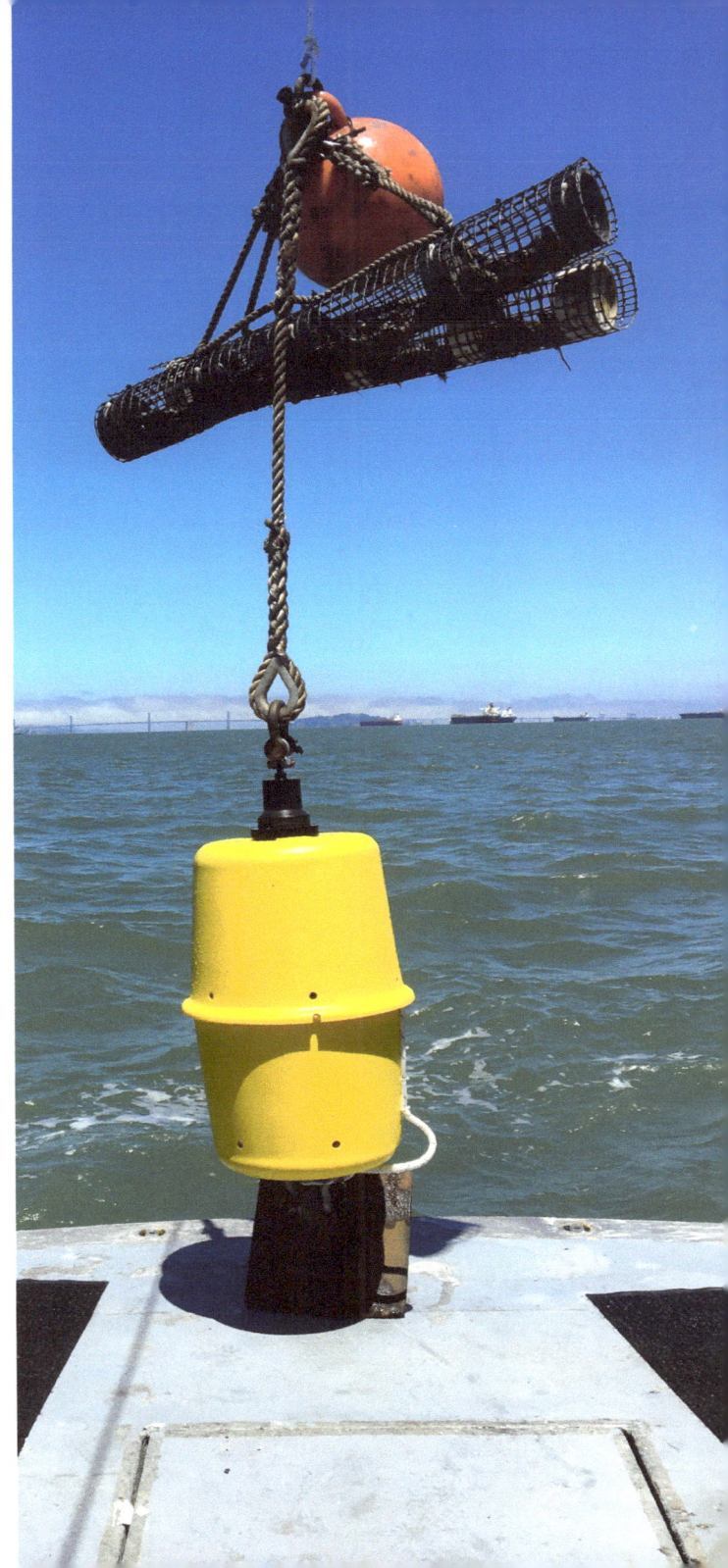

Technical Reports

Alternative Flame Retardants in San Francisco Bay: Synthesis and Strategy. Lin, D., and Sutton, R. 2018. https://www.sfei.org/documents/alternative-flame-retardants-san-francisco-bay-synthesis-and-strategy

Assessing the Impact of Periodic Dredging on Macroinvertebrate-Prey Availability for Benthic Foraging Fishes: Final Study Plan and Preliminary Pilot Study Results. De La Cruz, S., Woo, I., Flanagan, A., and Mittelstaedt, H. 2017. http://www.sfei.org/documents/assessing-impact-periodic-dredging-macroinvertebrate-prey-availability-benthic-foraging

Characterization of Sediment Contamination in Central Bay Margin Areas. Yee, D., Wong, A., Shimabuku, I., and Trowbridge, P. 2017. http://www.sfei.org/documents/characterization-sediment-contamination-central-bay-margin-areas-0

Conceptual Model to Support PCB Management and Monitoring in the Emeryville Crescent Priority Margin Unit. Davis, J.A., Yee, D., Gilbreath, A.N., and McKee, L.J. 2017. http://www.sfei.org/documents/conceptual-model-support-pcb-management-and-monitoring-emeryville-crscent-priority-margin

Conceptual Model to Support PCB Management and Monitoring in the San Leandro Bay Priority Margin Unit: Phase I. Yee, D., Gilbreath, A.N., McKee, L.J., and Davis, J.A. 2017. http://www.sfei.org/documents/conceptual-model-support-pcb-management-and-monitoring-san-leandro-bay-priority-margin

Conceptual Model to Support PCB Management and Monitoring in the San Leandro Bay Priority Margin Unit: Phase III. Yee, D.; Gilbreath, A.; McKee, L.; Davis, J. 2018. https://www.sfei.org/documents/conceptual-model-support-pcb-management-and-monitoring-san-leandro-bay-priority-margin-0

Contaminant Concentrations in Sport Fish from San Francisco Bay, 2014. Sun, J., Davis, J.A., Bezalel, S.N., Ross, J.R.M., Wong, A., Fairey, R., Bonnema, A., Crane, D.B., Grace, R., and Mayfield, R. 2017. http://www.sfei.org/documents/contaminant-concentrations-sport-fish-san-francisco-bay-2014

Current Knowledge and Data Needs for Dioxins in San Francisco Bay. Yee, D., Wong, A., and Hetzel, F. 2018. https://www.sfei.org/documents/dioxin-synthesis

The effects of kaolin clay on the amphipod Eohaustorius estuarius: Part Two. Anderson, B., Phillips, B., and Voorhees, J. 2017. http://www.sfei.org/documents/effects-kaolin-clay-amphipod-eohaustorius-estuaries-part-two

Estrogen Receptor In Vitro Assay Linkage Studies. Denslow, N., Kroll, K., Mehinto, A., and Maruya, K. 2018. https://www.sfei.org/documents/estrogen-receptor-vitro-assay-linkage-studies

Guadalupe River mercury concentrations and loads during the large rare January 2017 storm. McKee, L.J., Gilbreath, A.N., Pearce, S.A., and Shimabuku, I., 2018. https://www.sfei.org/documents/guadalupe-river-mercury-concentrations-and-loads-during-large-rare-january-2017-storm

Gut Contents Analysis of Four Fish Species Collected in the San Leandro Bay RMP PCB Study in August 2016. Jahn, A. 2018. https://www.sfei.org/documents/gut-contents-analysis-four-fish-species-collected-san-leandro-bay-rmp-pcb-study-august

Linkage of In Vitro Assay Results With In Vivo Endpoints. Phases 1 and 2. Denslow, N., Kroll, K., Jayasinghe, S., Adeyemo, O., Lavelle, C., Li, E., Mehinto, A.C., Bay, S., and Maruya, K. 2017. http://www.sfei.org/documents/linkage-vitro-assay-results-vivo-endpoints

Per and Polyfluoroalkyl Substances (PFASs) in San Francisco Bay: Synthesis and Strategy. Sedlak, M., Sutton, R., Wong, A., and Lin, D. 2018. https://www.sfei.org/documents/and-polyfluoroalkyl-substances-pfass-san-francisco-bay-synthesis-and-strategy

Pollutants of concern reconnaissance monitoring final progress report, water years 2015, 2016, and 2017. Gilbreath, A.N., Wu, J., Hunt, J.A., and McKee, L.J. 2018. https://www.sfei.org/documents/pollutants-concern-reconnaissance-monitoring-water-years-2015-2016-and-2017

Regional Watershed Spreadsheet Model (RWSM): Year 6 Progress Report. Wu, J., Gilbreath, A.N., and McKee, L.J. 2017. http://www.sfei.org/documents/regional-watershed-spreadsheet-model-rwsm-year-6-final-report

San Francisco Bay California Toxics Rule Priority Pollutant Ambient Water Monitoring Report. Yee, D., and Ross, J. 2017. http://www.sfei.org/documents/san-francisco-bay-california-toxics-rule-priority-pollutant-ambient-water-monitoring

San Francisco Bay Triennial Bird Egg Monitoring Program for Contaminants - 2016 Data Summary. Ackerman, J., Hartman, A., Herzog, M.P., and Toney, M. 2016. http://www.sfei.org/documents/san-francisco-bay-triennial-bird-egg-monitoring-program-contaminants-2016-data-summary

San Leandro Bay PCB Study Data Report. Davis, J., Yee, D., Fairey, R., and Sigala, M. 2017. http://www.sfei.org/documents/san-leandro-bay-pcb-study-data-report

Sediment Supply to San Francisco Bay, Water Years 1995 through 2016: Data, trends, and monitoring recommendations to support decisions about water quality, tidal wetlands, and resilience to sea level rise. Schoellhamer, D., McKee, L., Pearce, S., Kauhanen, P., Salomon, M., Dusterhoff, S., Grenier, L., Marineau, M., and Trowbridge P. 2018. https://www.sfei.org/documents/sediment-supply-san-francisco-bay

Selenium in Muscle Plugs of White Sturgeon from North San Francisco Bay, 2015-2017. Sun, J., Davis, J., and Stewart, R. 2018. https://www.sfei.org/documents/selenium-muscle-plugs-white-sturgeon-north-san-francisco-bay-2015-2017

Selenium in White Sturgeon from North San Francisco Bay: The 2015-2017 Sturgeon Derby Study. Sun, J., David, J., Stewart, R., and Palace, V. 2018. https://www.sfei.org/documents/selenium-white-sturgeon-north-san-francisco-bay-2015-2017-sturgeon-derby-study

North Bay Selenium Monitoring Design. Grieb, T., Roy, S., Rath, J., Stewart, R., Sun, J., and Davis, J. 2018. https://www.sfei.org/documents/north-bay-selenium-monitoring-design

Statistical Methods Development and Sampling Design Optimization to Support Trends Analysis for Loads of Polychlorinated Biphenyls from the Guadalupe River in San Jose, California, USA, Final Report. Melwani, A.R., Yee, D., McKee, L., Gilbreath, A., Trowbridge, P., and Davis, J.A. 2018. https://www.sfei.org/documents/statistical-methods-development-and-sampling-design-optimization-support-trends-analysis

Water and Suspended-Sediment Flux Measurements at the Golden Gate, 2016-2017. Downing-Kunz, M., Schoellhamer, D., and Work, P. 2017. https://www.sfei.org/documents/water-and-suspended-sediment-flux-measurements-golden-gate-2016-2017#

Water Column Selenium Concentrations in the San Francisco Bay-Delta: Recent Data and Recommendations for Future Monitoring. Chen, L., Roy, S., Rath, J., and Grieb, T. 2017. http://www.sfei.org/documents/water-column-selenium-concentrations-san-francisco-bay-delta-recent-data-and

Nutrient Management Strategy Products

Nutrient Management Strategy Science Program: FY2017 Annual Report. Holleman, R., MacVean, L., Mckibben, M., Sylvester, Z., Wren, I., and Senn, D. 2017. https://www.sfei.org/documents/nutrient-management-strategy-science-program

Synthesis of Current Science: Influence of Nutrient Forms and Ratios on Phytoplankton Production and Community Composition. Senn et al. 2016. https://www.sfei.org/documents/synthesis-current-science-influence-nutrient-forms-and-ratios-phytoplankton-production-and

San Francisco Bay Nutrient Management Strategy Science Plan. Final Approved Plan, September 2016. Senn, D., and Novick, E. 2016. https://www.sfei.org/documents/nutrient-management-strategy-science-plan-report

San Francisco Bay Nutrient Management Strategy Observation Program. Senn, D., and Trowbridge, P. 2016. https://www.sfei.org/documents/san-francisco-bay-nutrient-management-strategy-observation-program

San Francisco Bay Interim Model Validation Report. Holleman, R., Nuss, E., and Senn, D. 2017. https://www.sfei.org/documents/san-francisco-bay-interim-model-validation-report

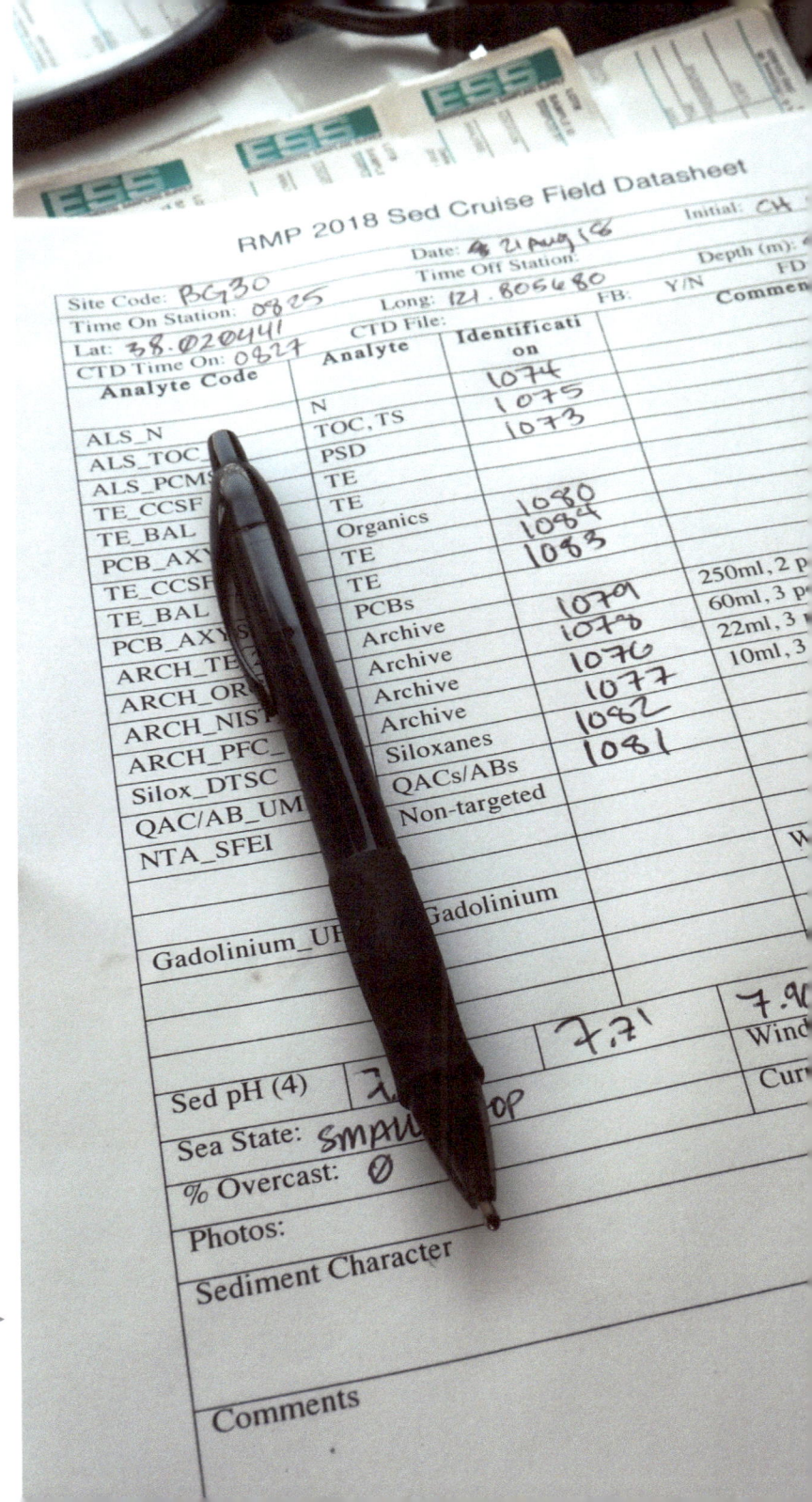

Sediment cruise field data sheet. Photograph by Shira Bezalel. ▶

Published Datasets

The RMP datasets that were finalized in the last two years are listed below. These datasets are available on cd3.sfei.org. To download the dataset, click on "Direct Download Tool" and select the associated Project Code and Analyte Group.

Dataset	Project Code
2017 RMP Pollutants of Concern Stormwater	STLS Monitoring RMP WY2017
2016 San Leandro Bay Sum of PCBs for Water and Sediment	RMP Special Study San Leandro Bay PCB Study (SEP) Select the "PCB" analyte group.
RMP Status and Trends Water Toxicity	2017 RMP Status and Trends Select the "'Conventional' or 'End Point" analyte group.
2002 and 2003 California Toxics Rule	2002 CTR Study' or '2003 CTR Study
2017 Margins Special Study (PCBs, Hg, MeHg, TEs, grain size, CHN, TOC, pH)	2017 RMP Margins Special Study
2010, 2011, 2015 and 2017 Field Measurements for RMP Water Cruises (DO, pH, salinity, temp, conductivity)	'RMP Status and Trends Project' with the year of interest
2015 and 2017 Margins Habitat	'2015 RMP Margins Special Study' or '2017 RMP Margins Special Study' Select the "Habitat" analyte group.
2017 RMP Water (MeHg, Cu, Se and Nutrients)	2017 RMP Status and Trends
RMP SEP: 2016 San Leandro Bay Fish	RMP Special Study San Leandro Bay PCB Study (SEP)
2003-2014 Historic Sportfish Dioxin Furan TEQs	'2003 RMP Fish' or '2006 RMP Fish' or '2009 RMP Fish' or '2014 RMP Fish' Select the "DioxinsDibenzofurans" analyte group and "TEQs" analyte sub group.
2017 Guadalupe River Flood Monitoring	STLS Monitoring WY2017 - Guadalupe River
San Leandro Bay Phase 3 Fish Tissue	RMP Special Study San Leandro Bay PCB Study (SEP)
2014 RMP Fish	2014 RMP FISH
2016 RMP Bird	2016 RMP EEPS Pilot Study
2016 RMP Bivalve	2016 RMP Status and Trends
RMP SEP: 2016 San Leandro Bay Water and Sediment	RMP Special Study San Leandro Bay PCB Study (SEP)

The 2017 water cruise. Photograph by Meg Sedlak. ▶

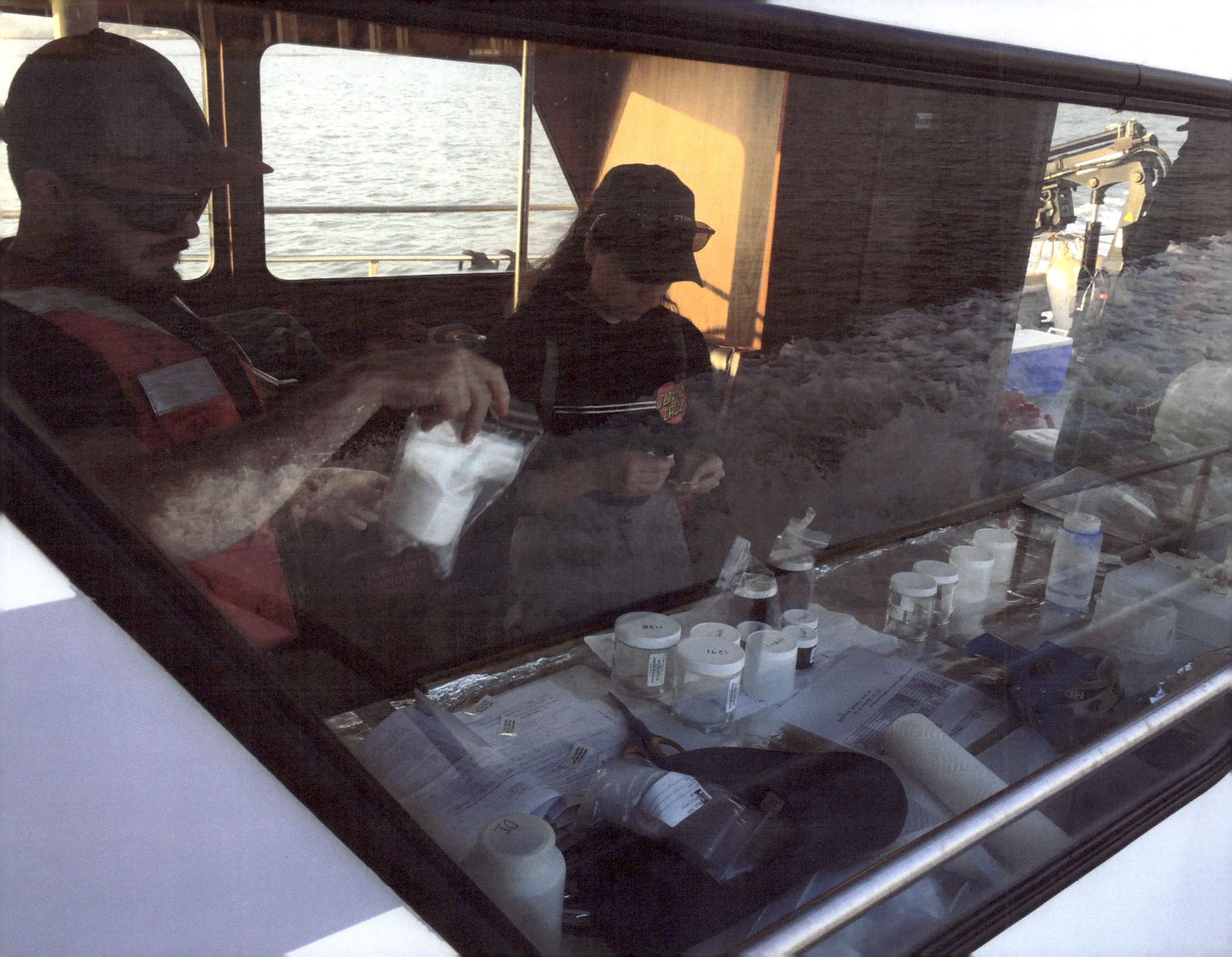

PROGRAM
AREA
UPDATES

STATUS AND TRENDS MONITORING

Background

The Status and Trends monitoring program is the core of the RMP's long-term monitoring strategy. Since the beginning of the RMP in 1993, water, sediment, and bivalve tissues have been monitored regularly in the open Bay. Sport fish and bird egg monitoring were added to the Program in 1997 and 2002, respectively.

Annual sampling of water and sediment had sufficiently documented trends and spatial patterns that varied by pollutant. This led to a revision between 2011 and 2014 to free up resources for special studies and other topics.

Sediment monitoring in the shallow margin areas of the Bay is currently being considered for addition to the Status and Trends program through a pilot study on Central and South Bay.

RELATION TO PERMIT REQUIREMENTS

NPDES Permits

- Receiving water compliance monitoring for NPDES discharge permit holders

- Provides data for Reasonable Potential Analyses

- Provides data for evaluating site specific objectives for copper and cyanide

Essential Fisheries Habitat Consultation, PCBs TMDL, Mercury TMDL

- Provides data to calculate dredged material testing thresholds in sediment and in-Bay disposal limits

USES OF RMP STATUS AND TRENDS DATA FOR MANAGEMENT DECISIONS

- Defining ambient conditions in the Bay

- Water Quality Assessment – 303(d) impairment listings or de-listings

- Determination of whether there is reasonable potential that a NPDES-permitted discharge may cause violation of a water quality standard

- Evaluation of water and sediment quality objectives

- Dredged material management

- Development and implementation of TMDLs for mercury, PCBs, and selenium

- Site-specific objectives and anti-degradation policies for copper and cyanide

- Development and evaluation of a Nutrient Assessment Framework (i.e., development of water quality objectives)

PRIORITY QUESTIONS

1 Are contaminants at levels of concern?

2 What are the concentrations and masses of priority contaminants in the Bay, its compartments, and its segments?

3 Are there particular regions of concern?

4 Have concentrations and masses increased or decreased?

STATUS AND TRENDS MONITORING

RECENT FINDINGS

In 2015, the full suite of 126 priority pollutants in the California Toxics Rule were monitored in RMP water samples. The concentrations of most contaminants were below detection. Those analytes that were detected had concentrations below their water quality criteria.

In 2015, the RMP monitored sediment in the margin areas of Central Bay. The study determined the ambient concentrations of PCBs, mercury, and other contaminants in these areas. On average, PCB concentrations were 4-5 times higher in the margins than in the open Bay. The study also detected a number of "warm spots" where the concentrations of contaminants were significantly elevated and one previously unknown "hot spot".

In 2017, the RMP published the latest information on contaminant concentrations in sport fish tissue. The most recent data show that there is no long-term trend for mercury and little evidence of PCB declines in important sport fish species.

Copper concentrations in water, last monitored in 2017, remain below trigger levels. However, a comparison of new and old lab methods in 2017 indicated an inconsistency which will be resolved for the 2019 sampling.

Over a decade of monitoring shows that PBDE levels have declined in bivalves, bird eggs, sport fish, and sediment following nationwide phase-outs and state bans of these toxic and persistent flame retardant chemicals. The RMP now considers PBDEs to be in the "low concern" category and will reduce, but not eliminate, monitoring for them.

WORKPLAN HIGHLIGHTS

Long-term monitoring of

- **water** biennially,
- **sediment** once every four years,
- **bivalves** biennially,
- **bird eggs** triennially, and
- **sport fish** once every five years

In 2017, the RMP monitored sediment in the margin areas of South Bay. A report summarizing the results of this study, compared to results from the margin areas of Central Bay, will be completed in 2018.

RMP partner laboratories participated in an intercalibration study organized by the Southern California Coastal Water Resources Project. The study will help the laboratories to produce comparable results and will help the RMP to have confidence that changes in status and trends indicators are due to changes in the Bay, not the labs. Final results will be reported in 2018.

In 2018, the RMP is conducting three types of status and trends monitoring: sediment, bivalves, and bird eggs. Data from this monitoring will be reported in the 2019 Pulse of the Bay.

Planning is already underway for the 2019 sport fish monitoring, which is a major effort that is completed once every 5 years.

Collecting a water sample. Photograph by Shira Bezalel.▶

COLLABORATORS

- San Francisco Bay Regional Water Quality Control Board
- US Environmental Protection Agency
- Applied Marine Sciences
- SGS-AXYS Analytical
- Brooks Analytical Labs
- City and County of San Francisco
- US Geological Survey
- ALS Environmental
- Pacific EcoRisk
- Moss Landing Marine Laboratory
- Marine Pollution Studies Laboratory
- Coastal Conservation & Research
- City of San Jose

EMERGING CONTAMINANTS

Background

Contaminants of emerging concern (CECs) are not regulated or routinely monitored, yet have the potential to enter the environment and harm people or wildlife. Through its focus on emerging contaminants, the RMP aims to identify problem chemicals before they harm Bay wildlife. The RMP's decades-long effort has made the Bay one of the most thoroughly studied estuaries in the world. Surveillance has identified four contaminants or classes of moderate concern:

- PFOS - a stain and water repellant
- PFOA and related long-chain perfluorocarboxylates - previously used in fire-fighting foams and to make non-stick coatings
- Fipronil - a widely used insecticide
- Alkylphenol and alkylphenol ethoxylates - a class of detergent ingredients

RELATION TO PERMIT REQUIREMENTS

- Municipal wastewater dischargers may opt into the alternate monitoring and reporting permit with fees that provide additional funds to support the RMP and its emerging contaminants monitoring.

- The revised Municipal Regional Stormwater Permit (2015) requires monitoring studies of key emerging contaminants, including flame retardants, PFOS and related compounds, and pesticides.

USES OF RMP EMERGING CONTAMINANT DATA FOR DECISIONS

- Regional Action Plans for emerging contaminants

- Early management intervention, including green chemistry and pollution prevention

- State and federal pesticide regulatory programs

PRIORITY QUESTIONS

1 Which CECs have the potential to adversely impact beneficial uses in San Francisco Bay?

2 What are the sources, pathways, and loadings leading to the presence of individual CECs or groups of CECs in the Bay?

3 What are the physical, chemical, and biological processes that may affect the transport and fate of individual CECs or groups of CECs in the Bay?

4 Have the concentrations of individual CECs or groups of CECs increased or decreased in the Bay?

5 Are the concentrations of individual CECs or groups of CECs predicted to increase or decrease in the future?

6 What are the effects of management actions?

EMERGING CONTAMINANTS

RECENT FINDINGS

The RMP completed the first major revision of its CEC Strategy document. The review affirmed the value of the RMP's three-element strategy: 1) CEC monitoring and risk evaluation; 2) reviewing literature and other programs to identify new analytes; and 3) using non-targeted techniques to scan for additional concerns. New management questions were added. Information regarding temporal trends was added to the tiered prioritization framework used to classify the level of concern associated with emerging contaminant compounds or classes. A revised multi-year plan and recommendations for Status and Trends monitoring were also provided.

PBDEs were downgraded from Moderate to Low Concern for the Bay. Status & Trends monitoring of key matrices will continue for at least two cycles.

Synthesis of extensive RMP monitoring data on poly- and perfluoroalkyl substances (PFASs) and consultation with international experts resulted in the recommendation that perfluorooctanoic acid (PFOA) and similar long-chain carboxylates be considered Moderate Concerns for San Francisco Bay. The strategy for future studies includes continued Status & Trends monitoring of sport fish and bird eggs, and Special Studies on sediment, harbor seals, and stormwater using advanced analytical techniques.

Results from non-targeted analysis of Bay water samples suggest stormwater-influenced sites may have a number of diverse, unique contaminants at relatively high abundances. Follow-up studies that target compounds linked to pollution from vehicles, roadways, and other outdoor urban sources are underway.

WORKPLAN HIGHLIGHTS

Launch of Multi-year Monitoring Effort for Emerging Contaminants in Stormwater: Findings from the RMP's recent non-targeted analysis has resulted in a new focus on unique and rarely studied contaminants derived from vehicles and roadways. A major effort to investigate these and other emerging contaminants in Bay Area stormwater begins this fall.

Non-targeted Analysis of Sediment: A broad scan of Bay margin (near-shore) and ambient sediment for both water-soluble (polar) and fat-soluble (nonpolar) compounds will be used to identify unexpected contaminants that may merit further investigation.

Non-targeted Analysis of Runoff Impacted by North Bay Fires: The RMP is supporting North Bay communities devastated by last fall's wildfires through an effort to scan post-fire stormwater runoff for CECs, complementing existing assessments for conventional contaminants. Use of novel non-targeted methods will improve our understanding of the risks associated with toxic contaminants linked to wildfires.

COLLABORATORS

- SGS AXYS
- Bay Area Clean Water Agencies
- California Department of Toxic Substances Control
- California Department of Pesticide Regulation
- Duke University
- San Francisco Bay Regional Water Quality Control Board
- San Diego State University
- Southern Illinois University
- Southern California Coastal Water Research Project
- TDC Environmental
- United States Geological Survey

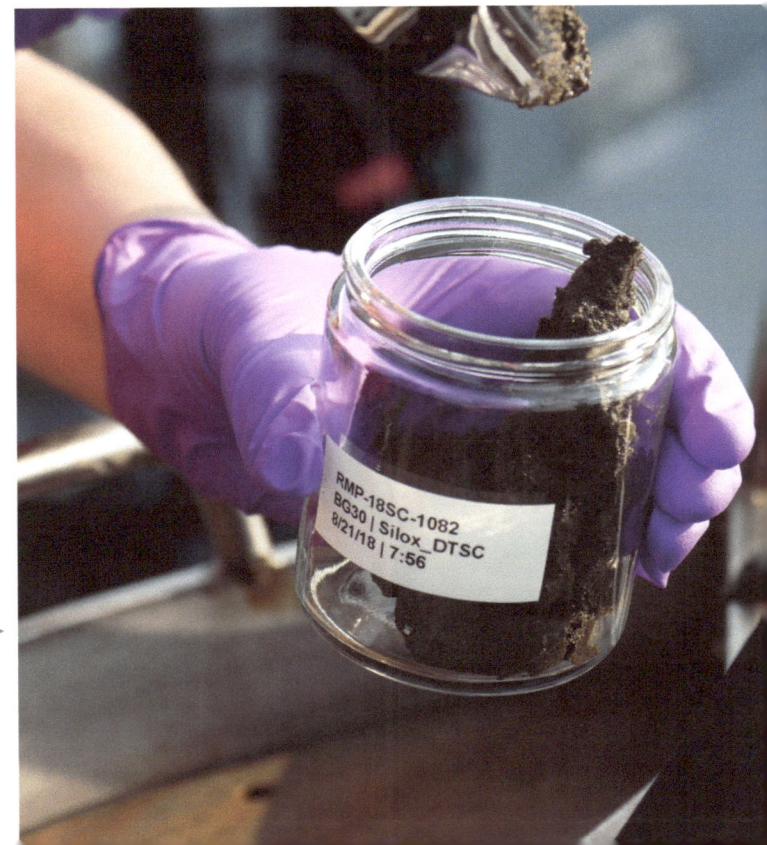

Filling a sample jar for CEC analysis. Photograph by Shira Bezalel. ▶

SMALL TRIBUTARY LOADING

NOTE: "Small tributary" refers to the rivers, creeks, and storm drains that enter the Bay from the nine counties that surround the Bay.

Background

San Francisco Bay PCB and mercury TMDLs were established to address health risks to humans and wildlife. Urban tributary loads are named in the TMDLs as the primary controllable source for reducing impairment. Other pollutants of concern (POCs) in urban stormwater include copper, PBDEs, nutrients, pesticides, and emerging contaminants.

To address information needs associated with improving understanding about these POCs, the Small Tributaries Loading Strategy (STLS), first written in 2009, was updated in 2018 to include a trends component to help prioritize and coordinate the activities of the RMP and Bay Area Storm-water Management Agencies Association permittees. STLS studies conducted over past decade have been focusing on locating, quantifying, and managing PCBs, mercury, and other pollutants in the urban environment to support management actions. Going forward, an increasing emphasis will be placed on tracking trends in loading and best management practice (BMP) implementation progress, through a combination of monitoring and modeling.

RELATION TO PERMIT REQUIREMENTS

Addresses monitoring requirements specified in the Municipal Regional Stormwater Permit

- Pollutants of Concern Monitoring
- Wet Weather Pesticides and Toxicity Monitoring
- Implement Control Measures to Achieve Mercury/PCB Load Reductions
- Assess Mercury/PCB Load Reductions from Stormwater
- Plan and Implement Green Infrastructure to Reduce Mercury/PCB loads
- Prepare Implementation Plan and Schedule to Achieve TMDL Allocations

USES OF RMP SMALL TRIBUTARY LOADING DATA FOR DECISIONS

- Refining pollutant loading estimates for future TMDL updates

- Informing provisions of the current and future versions of the Municipal Regional Stormwater Permit (MRP)

- Identifying small tributaries to prioritize for management actions

- Informing decisions on the best management practices for reducing pollutant concentrations and loads

- Tracking effectiveness of load reduction in individual small tributaries

PRIORITY QUESTIONS

1 What are the loads or concentrations of pollutants of concern from small tributaries to the Bay?

2 Which are the "high-leverage" small tributaries that contribute or potentially contribute most to Bay impairment by pollutants of concern?

3 How are loads or concentrations of pollutants of concern from small tributaries changing on a decadal scale?

4 Which sources or watershed source areas provide the greatest opportunities for reductions of pollutants of concern in urban stormwater runoff?

5 What are the measured and projected impacts of management action(s) on loads or concentrations of pollutants of concern from the small tributaries, and what management action(s) should be implemented in the region to have the greatest impact?

SMALL TRIBUTARY LOADING

RECENT FINDINGS

Based on winter storm sampling by the RMP and the Bay Area Stormwater Management Agencies Association (BASMAA) to date, watersheds with the highest PCB concentrations on particles are Pulgas Creek Pump Station in San Carlos, a ditch on Industrial Rd. in San Carlos, Line 12H at Coliseum Way in Oakland, Santa Fe Channel in Richmond, Pulgas Pump Station-North in San Carlos, Gull Drive storm drain in South San Francisco, and an outfall at Gilman Street in Berkeley. The outfall at Gilman Street and the Santa Fe Channel sites also appear to have relatively high concentrations of mercury.

Remote sediment sampler testing at 14 sites is now complete and proven as a useful lower-cost stormwater characterization tool, especially for PCBs. These samplers will be used for characterizing stormwater concentrations on particles at new sites in water year 2019.

A statistical model was developed for the Guadalupe River based on a turbidity-PCB relationship and additional climatic, seasonal, and inter-annual factors. Power analysis using the model indicated that a discrete-based sampling of 4 storms every other year could detect PCB load declines of 25% over a 20-year period, in contrast to composite-based sampling that could only detect declines of 75%. No significant trend was detected for the period of record (2003-2014).

A rare five-year storm event was sampled in the Guadalupe River in January 2017. The load measured during the storm event was 70 kg, far more than the total wet season loads for every year since 2003, the last time a storm of this magnitude occurred in the system. Sediment load for the wet season was the largest measured to date for this system.

WORKPLAN HIGHLIGHTS

Stormwater Reconnaissance Sampling: Over the past four years, the RMP, in collaboration with BASMAA member agencies, has funded a watershed characterization reconnaissance study to support a weight-of-evidence approach for the identification and management of PCB and mercury sources. This effort will continue, providing data on concentrations in water and on sediment particles to identify high-leverage watersheds and subwatersheds within larger areas of older urban and industrial land use. To decrease costs and increase ease of data collection, remote sampler methods will be used in new sample locations while manual composite sampling methods will be primarily used to revisit previously sampled locations.

Advanced Data Analysis: Reconnaissance data collected during single storms has provided good evidence to support enhanced management effort in watersheds with high PCB concentrations in water and on sediment particles. This effort will mine and analyze all the existing stormwater data with the primary goals of developing an improved method to identify and rank watersheds of management interest for further investigation, including watersheds with moderate or low concentrations, as well as to guide future sampling design.

Trends Strategy: The evaluation of stormwater loading trends in relation to management efforts and beneficial use impacts is an important new focus. To support this focus, the 2016 STLS trends strategy was updated in 2018 and expanded to outline a multi-year plan to make an assessment of trends in regional PCB loads over decadal scales through a combination of modeling and monitoring. The first step of implementing this multi-year plan is to develop a detailed Modeling Implementation Plan with input and oversight by STLS and the SPLWG to guide regional modeling efforts over the next few years.

▲ Collecting a stormwater sample. Photograph by Meg Sedlak.

NUTRIENTS

Background

San Francisco Bay receives some of the highest nitrogen loads among estuaries worldwide, yet has not historically experienced the water quality problems typical of other nutrient-enriched estuaries. It is not known whether this level of nitrogen loading, which will continue to rise in proportion to human population increase, is sustainable over the long term. Special studies and expanded monitoring carried out through the Nutrient Management Strategy have revealed some water quality conditions that have been associated with nutrient over-enrichment in other estuaries (e.g., recurring low dissolved oxygen in some margin habitats and consistent detection of multiple toxins produced by harmful algae). Potential impacts of these conditions on human and ecological health need to be more extensively evaluated and causal factors determined. A further complication is that the Bay's response to nutrients is influenced by many physical and biological factors including suspended sediment concentrations, light availability, freshwater inputs, and ocean conditions. These factors themselves vary by Bay sub-embayment and due to regional land and water management and climate oscillations. Therefore, a wide range of monitoring and special studies is needed to understand what might happen to Bay water quality as a result of changes in nutrients and other factors.

RELATION TO PERMIT REQUIREMENTS

The Bay-wide nutrient permit that went into effect in 2014 includes a provision to support science and monitoring to inform future permitting decisions.

USES OF RMP NUTRIENT DATA FOR DECISIONS

- Development of nutrient numeric endpoints and assessment framework

- Evaluating need for revised objectives for dissolved oxygen and other parameters

- Assessment of water quality impairment status

- Implementation of NPDES permits for wastewater and stormwater

PRIORITY QUESTIONS

1 What conditions in different Bay habitats would indicate that beneficial uses are being protected versus experiencing nutrient-related impairment?

2 In which subembayments or habitats are beneficial uses being supported? Which subembayments or habitats are experiencing nutrient-related impairment?

3 A. To what extent is nutrient over-enrichment, versus other factors, responsible for current impairments?
B. What management actions would be required to mitigate those impairments and protect beneficial uses?

4 A. Under what future scenarios could nutrient-related impairments occur, and which of these scenarios warrant pre-emptive management actions?
B. What management actions would be required to protect beneficial uses under those scenarios?

5 What nutrient sources contribute to elevated nutrient concentrations in subembayments or habitats that are currently impaired, or would be impaired in the future, by nutrients?

6 When nutrients exit the Bay through the Golden Gate, where are they transported and how do they influence water quality in the Gulf of Farallones or other coastal areas?

7 What specific management actions, including load reductions, are needed to mitigate or prevent current or future impairment?

NUTRIENTS

RECENT FINDINGS

In 2016, the Nutrient Management Strategy finished a 10-year Science Plan for addressing monitoring and research needs for the complicated issue of nutrients in the Bay.

High-frequency sensors are providing new data for identifying the mechanisms that drive dissolved oxygen concentrations in the Bay, such as algae blooms, tidal currents, suspended sediment, and stratification of the water column that limits transfer of oxygen to the bottom waters.

Studies conducted to date indicate that algae growth is most often limited by factors other than nutrients, such as high turbidity and strong tidal mixing, but the role of nutrients in fueling algae blooms at certain times and locations still needs to be resolved.

Algae that produce potent toxins have been detected in the Bay and these toxins are regularly detected in water and shellfish at levels that justify continued investigation.

Major progress on numerical models has been made in the first two years of the program. A major validation report was produced in 2017 which showed that the hydrodynamic model in its current state has sufficient skill in representing transport in South Bay to support water quality studies with a South Bay focus.

WORKPLAN HIGHLIGHTS

Data from high-frequency sensors and fish trawls in Lower South Bay are being synthesized to explore the issue of where and when there is adequate dissolved oxygen to support resident fish species. The report, which will be completed in 2018, is a collaboration between SFEI and the University of California Davis.

Data on harmful algae and their toxins in the Bay from 1993 to 2014 will be summarized in a synthesis report in 2018. This report will leverage 20 years of shipboard monitoring of the main channel of San Francisco Bay to investigate whether there is evidence that harmful algae reside, persist, or bloom within the Bay.

Ongoing and future modeling efforts include adding algae growth calculations and expanding the range of the models into the Delta and the sloughs of Lower South Bay.

Funding for a Supplemental Environmental Project is being used for a major study on harmful algae and toxins. The study will investigate whether toxins are accumulating in small fish and mussels. The use of new molecular techniques to identify harmful algae will also be tested. A report on this study will be prepared in 2019.

COLLABORATORS

- San Francisco Bay Regional Water Quality Control Board
- Bay Area Clean Water Agencies
- Additional members of the Nutrient Management Strategy Steering Committee
- US Geological Survey – Sacramento
- US Geological Survey – Menlo Park
- UC Santa Cruz
- Deltares
- UC Berkeley
- San Francisco State University
- US Environmental Protection Agency

A multi-sensor water quality probe. Photograph by Zephyr Sylvester. ▶

SELENIUM

Background

San Francisco Bay has been listed as impaired by selenium since 1990. Although water concentrations are below water quality thresholds, several wildlife species may be at risk for selenium toxicity. White sturgeon, a benthic species, is recognized as a key indicator of selenium impairment in the Bay due to its susceptibility to selenium bioaccumulation.

In 2016, a TMDL for North Bay was approved, establishing numerical selenium targets for water and white sturgeon tissue. In addition, USEPA proposed criteria for selenium in Bay-Delta fish, clams, and water in June 2016.

The RMP Selenium Workgroup was established in 2014 to develop monitoring strategies to inform implementation of the North Bay TMDL and consideration of a TMDL for the South Bay. Selenium in water, sediment, and tissue are regularly monitored through the Status and Trends program. Recent special studies have focused on developing tissue monitoring methods for white sturgeon. A monitoring plan for a suite of indicators (including water, clams, and sturgeon) that can provide an early indication of changing selenium exposure in the North Bay was developed in 2018.

RELATION TO PERMIT REQUIREMENTS

Supports the development and implementation of selenium TMDLs for North and possibly South Bay, as well as USEPA site-specific selenium criteria for the San Francisco Bay-Delta.

USES OF RMP SELENIUM DATA FOR DECISIONS

- North Bay Selenium TMDL

- Proposed USEPA Selenium Criteria for the Bay-Delta

- South Bay Selenium TMDL (under consideration)

PRIORITY QUESTIONS

General

1 What are appropriate thresholds?

2 Are the beneficial uses of San Francisco Bay impaired by selenium?

3 What is the spatial pattern of selenium impairment?

4 How do selenium concentrations and loadings change over time?

5 What is the relative importance of each pathway of selenium loading in the Bay?

North Bay

6 Are the beneficial uses of north San Francisco Bay impaired by selenium?

7 Are changes occurring in selenium concentrations that warrant changes in management actions?

8 Will proposed changes in water flows and/or selenium loads in the Bay or upstream cause impairment in the North Bay?

SELENIUM

RECENT FINDINGS

Non-lethal monitoring of selenium in muscle plugs from white sturgeon (the species and tissue established as the impairment indicator in the TMDL) in 2015-2017 found that concentrations were significantly lower during the high flows of 2017 relative to the two prior drought years, confirming a pattern that was expected based on long-term monitoring of the clams that are a primary component of the sturgeon diet. Median concentrations in 2015 (10.9 ug/g dry weight) and 2016 (10.6 ug/g) approached the TMDL target of 11.3 ug/g, with 47% and 44% of individual samples above the target. In contrast, in the wet year of 2017 the median was 6.8 ug/g, and only 12% of the samples exceeded the target.

Sturgeon monitoring conducted in coordination with an annual sturgeon fishing derby in the western Delta, also in 2015-2017, showed that selenium concentrations in muscle are correlated with concentrations in ovaries and liver (tissues that are more closely linked to fish health risk), and that concentrations in muscle plugs are well-correlated with concentrations in muscle fillets.

A selenium monitoring design for the North Bay has been developed, with an emphasis on early detection of changes that could warrant changes in management approaches. The design was based on a review of available data and power analyses for water, clams, and sturgeon.

WORKPLAN HIGHLIGHTS

Continued monitoring of selenium in North Bay sturgeon muscle plugs on a biennial basis.

Implementation of the integrated monitoring design for Suisun Bay that includes water, clams, and sturgeon.

Evaluation of information needs for management of selenium in the South Bay.

COLLABORATORS

- California Department of Fish and Wildlife
- US Geological Survey - Menlo Park
- Tetra Tech
- US Environmental Protection Agency

Measuring the volume of a water sample. Photograph by Shira Bezalel.

PCBs

Background

PCB contamination is a high priority for Bay water quality managers due to concerns for risks to humans and wildlife. A TMDL was approved in 2009. Monitoring of small fish along the margins of the Bay in 2010 showed higher PCB concentrations than in the open Bay. In 2014, the RMP completed a synthesis report summarizing advances in understanding of PCBs in the Bay since the data synthesis for the PCBs TMDL.

An updated conceptual model presented in that report called for monitoring and management to focus on contaminated areas on the Bay margins. Local-scale actions within a margin area, or in upstream watersheds, will be needed to reduce exposure within that area. The multi-year workplan for PCBs is focusing on supporting a planned revision of the PCBs TMDL in 2020 by preparing to detect improvements in Bay high-priority margin areas in response to anticipated stormwater load reductions.

Site-specific conceptual models have been developed for two margin areas that are high priorities for water quality managers: the Emeryville Crescent and San Leandro Bay. A site-specific conceptual model for a third area — Steinberger Slough — is also under development.

RELATION TO PERMIT REQUIREMENTS

Addresses critical information needs identified in the PCB TMDL to be addressed by municipal and industrial wastewater dischargers and stormwater management agencies.

Addresses a requirement in the Municipal Regional Stormwater Permit: Fate and transport study of PCBs - Urban runoff impact on San Francisco Bay margins

USES OF RMP PCB DATA FOR DECISIONS

- PCBs TMDL and potential update

- Implementation of NPDES permits, including the Municipal Regional Permit for Stormwater

- Selecting management actions for reducing PCB impairment

- Updating the fish consumption advisory

PRIORITY QUESTIONS

1 What are the rates of recovery of the Bay, its segments, and in-Bay contaminated sites from PCB contamination?

2 What are the present loads and long-term trends in loading from each of the major pathways?

3 What role do in-Bay contaminated sites play in segment-scale recovery rates?

4 Which small tributaries and contaminated margin sites are the highest priorities for cleanup?

5 What management actions have the greatest potential for accelerating recovery or reducing exposure?

6 What are the near-term effects of management actions on the potential for adverse impacts on humans and aquatic life due to Bay contamination?

PCBs

COLLABORATORS

- Moss Landing Marine Laboratory
- AXYS Analytical
- Stanford University

RECENT FINDINGS

Shiner surfperch have a Bay-wide average concentration 9 times higher than the TMDL target, and these concentrations have resulted in an advisory from the Office of Environmental Health Hazards Assessment (OEHHA) recommending no consumption for all surfperch in the Bay. Concentrations in shiner surfperch and white croaker show no clear sign of decline.

An assessment of Emeryville Crescent established a conceptual model as a foundation for monitoring response to load reductions and for planning management actions. The key finding was that PCB concentrations in sediment and the food web could potentially decline fairly quickly (within 10 years) in response to load reductions from the watershed.

A conceptual model for PCBs in San Leandro Bay was completed in 2018. A simple mass budget suggested that San Leandro Bay should respond to reductions in watershed loads, but sediment concentrations have not declined since 1998, suggesting that continuing inputs are slowing recovery. Significant cleanup actions that have been recently completed or are happening soon on highly contaminated properties adjacent to San Leandro Bay should promote recovery.

WORKPLAN HIGHLIGHTS

Completion of the conceptual site model for Steinberger Slough in late 2018.

Field studies to address critical information gaps in San Leandro Bay and the Emeryville Crescent.

Baseline monitoring of four priority margin areas (Emeryville Crescent, San Leandro Bay, Steinberger Slough, and Richmond Harbor), beginning with shiner surfperch monitoring in 2019 (in coordination with the 2019 Status and Trends sport fish monitoring).

Scooping a sediment sample out of the Van Veen grab sampler. ▶
Photograph by Shira Bezalel.

EXPOSURE AND EFFECTS

Background

Studies under the RMP Exposure and Effects Workgroup began in 2001. The primary goal of early work was to identify indicator species to measure contaminant risks to wildlife and ecosystem health at various trophic levels, spatial scales, and levels of biological organization (e.g., cellular, organism, or population). Ongoing studies focus on developing tests and metrics to measure the potential toxic effects of contamination on aquatic life. The Exposure and Effects Workgroup also supports the Status and Trends and Emerging Contaminants components of the RMP.

RELATION TO PERMIT REQUIREMENTS

Essential Fisheries Habitat Consultation, PCBs TMDL, Mercury TMDL

Provides information and tools for setting dredged material testing thresholds and in-Bay disposal limits

USES OF RMP EXPOSURE AND EFFECTS DATA FOR DECISIONS

- Implementation of narrative water quality objectives for toxicity, bioaccumulation, and aquatic species populations and community ecology

- Implementation of sediment quality objectives

- Permitting decisions regarding dredging projects

- Contaminated sediment 303(d) listing and delisting decisions

PRIORITY QUESTIONS

1 What are the spatial and temporal patterns of impacts of sediment contamination?

2 Is chemical contamination the cause of observed sediment toxicity in the Bay?

3 What are the best tools to predict ecological effects from chemical contamination of sediments in the Bay?

4 Should any sediment contamination hotspots on the 303D list be de-listed?

5 Do spatial patterns in bioaccumulation in birds indicate particular regions of concern?

6 Are there any indications of ecological effects caused by exposure to specific chemicals or mixtures of contaminants in the Bay? (Overlap with ECWG)

7 What are acceptable levels of chemicals in dredged material for placement in the Bay, baylands, or restoration projects? (Overlap with Sediment WG)

8 Are there effects on fish, benthic species, and submerged habitats from dredging or placement of dredged material? (Overlap with Sediment WG)

EXPOSURE AND EFFECTS

RECENT FINDINGS

SQO analyses do not indicate severe impacts to benthos. SQO analyses of 125 RMP sites from 2008 to 2012 indicate that severe impacts to the benthic community are not observed in the Bay. Forty percent of the Bay was classified as Possibly Impacted, indicating that the impacts are small or uncertain due to conflicting lines of evidence. Laboratory studies by the University of California Davis showed that clay has size-specific mortality effects on the amphipod species used in RMP sediment toxicity testing. Larger amphipods appear to be more sensitive to clay particles. In 2016, additional studies using sediment samples from the Bay confirmed these findings. Future RMP sediment toxicity testing will focus on smaller amphipods to minimize this effect.

The University of Florida and the Southern California Coastal Water Resources Project demonstrated linkages between in vitro assays measuring estrogenic activity and in vivo endpoints in fish that point to population level effects. As a result of the study, the estrogenic response "bioanalytical tool" can be used to measure cumulate estrogenic activity from mixtures of chemicals in the Bay.

The US Geological Survey completed a pilot study on whether benthic habitat for foraging fish is affected by periodic dredging. The RMP contributed to the first phase of the study, which collected baseline information and did a power analysis to plan the full study. They found that mean macroinvertebrate density was greater in undredged reference areas than in dredged areas at four of five study sites. The pilot study also confirmed that a study design with 200 sediment cores would be able to detect a 50% difference in benthic species.

Measuring sediment pH in the Van Veen grab sampler. ▶
Photograph by Shira Bezalel.

WORKPLAN HIGHLIGHTS

The Southern California Coastal Water Resources Project is developing a benthic habitat index designed to work in multiple estuarine habitats across the United States, including the polyhaline, mesohaline, and oligohaline habitats in San Francisco Bay. The first phase of the study will be completed in early 2019.

A study is underway to develop a standard set of toxicity reference values (TRVs), which can be used in assessment of bioaccumulation testing results for dredged sediments. This study will save dredgers and regulators time and money by avoiding the need to conduct individual studies to develop TRVs, and also improve the efficiency and consistency of dredging project evaluations. The report from this study will be completed in 2018.

COLLABORATORS

- Southern California Coastal Water Research Project
- UC Davis – Granite Canyon
- University of Florida
- US Geological Survey – Western Ecological Research Center
- US Army Corps of Engineers
- Bay Conservation and Development Commission

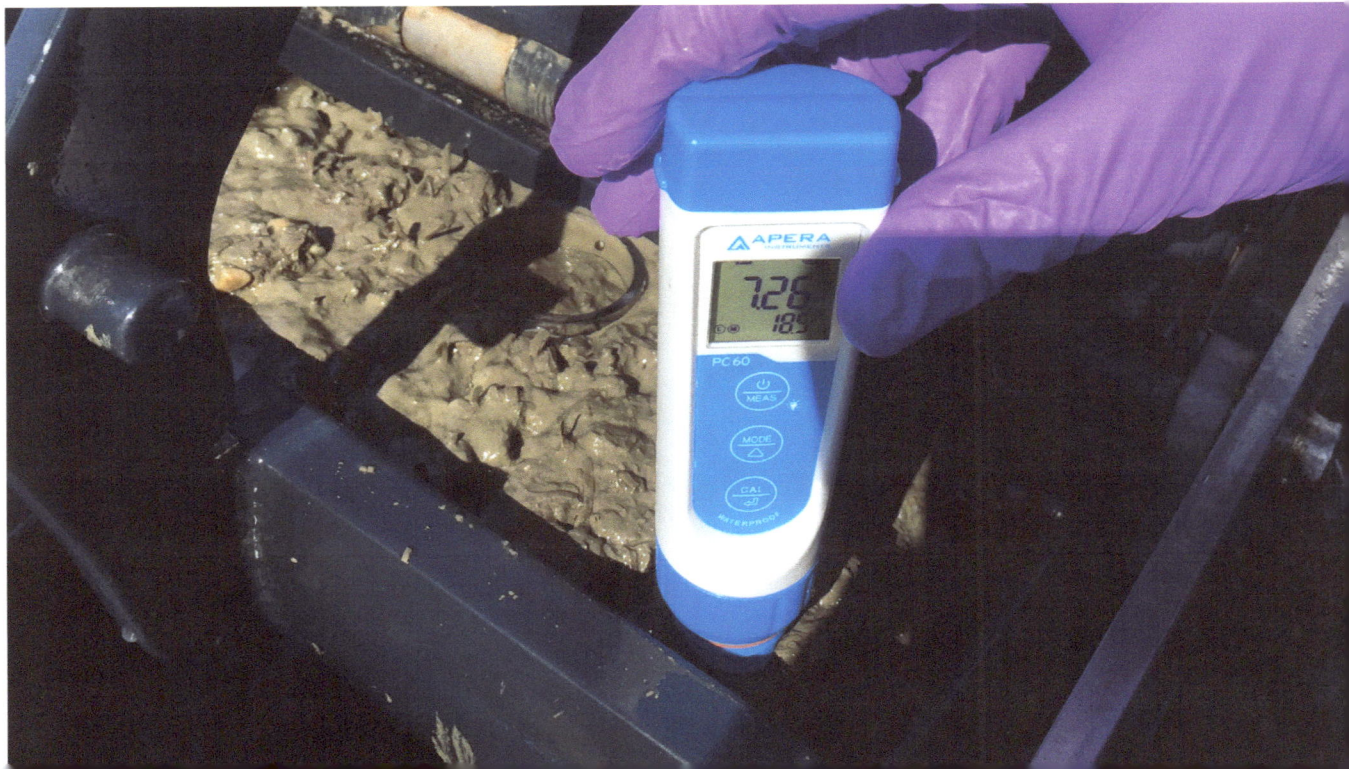

DIOXINS

Background

San Francisco Bay was placed on the State of California's 303(d) list of impaired waterways in 1998 as a result of elevated concentrations of dioxins and furans (commonly referred to as only 'dioxin') in fish. RMP studies of contaminants in Bay sport fish conducted since 1994 have found that dioxin concentrations have remained relatively unchanged over this time period and, in some species, continue to exceed a screening value for human consumption. Our understanding of dioxin in the Bay has improved due to special studies conducted over the past decade. Although the available information suggests progress will be slow toward Bay-wide reductions in concentrations in fish and resulting health risks to humans and wildlife, similar to PCBs, there may be localized opportunities to effect change at select, more highly impacted, sites.

RELATION TO PERMIT REQUIREMENTS

The Dioxin Strategy is generating the information needed to support development of appropriate effluent limits for municipal and industrial discharges. In addition, the information generated on ambient sediment concentrations is useful for establishing appropriate thresholds for in-Bay disposal and re-use of dredged sediments.

USES OF RMP DIOXIN DATA FOR DECISIONS

- Review 303(d) listings and establish TMDL development plan or alternative

- Defining ambient conditions in the Bay

- Evaluation of water and sediment quality objectives

- Evaluate trends in Bay conditions

- Dredged material management (in-Bay disposal thresholds, beneficial re-use opportunities)

- Estimate loads via various pathways for potential management

PRIORITY QUESTIONS

1 Are beneficial uses of SF Bay impaired by dioxins?

2 What is the dioxin reservoir in Bay sediment and water?

3 Have dioxin loadings or concentrations changed over time?

4 What is the relative contribution of each loading pathway as a cause of dioxin impairment in the Bay?

DIOXINS

RECENT FINDINGS

The key sport fish indicator species (shiner surfperch and white croaker) have tissue levels higher than the Water Board screening value of 0.14 ppt and show no sign of decline, but there is a great deal of uncertainty regarding the human health risk associated with dioxins in sport fish.

Dioxin toxic equivalents in Least Tern, Caspian Tern, and Forster's Tern eggs are at or above estimated thresholds for adverse effects; risks are especially significant in combination with dioxin-like PCBs.

Wetland sediment cores suggest rapidly declining inputs from local watersheds during recent decades, although additional coring data are needed to support this hypothesis.

Recent RMP monitoring of open Bay water and sediment did not show patterns suggesting large localized sources of dioxins in different areas of the Bay. However, data compiled from dredging projects suggest areas with higher concentrations due to local urban sources and limited transport and dispersion from near-shore areas.

Revised estimates of atmospheric deposition derived from local air monitoring (conducted by the California Air Resources Board) and monitoring of runoff loads from local watersheds (conducted by RMP and BASMAA Agencies) suggest about three-fold higher loads than estimated in the 2005 Conceptual Model Impairment Assessment. Although the CARB data are over 10 years old, and subsequent declines may have occurred, these load estimates may suggest higher steady state ambient concentrations and slower declines than previously projected for dioxins in the Bay.

Sediment in the Van Veen grab sampler. Photograph by Shira Bezalel. ▶

WORKPLAN HIGHLIGHTS

Work in 2017-2018 to summarize and synthesize results of monitoring conducted between 2008 and 2014 in response to information needs prioritized by the RMP Dioxin Workgroup is near completion, with a technical report completed in the 3rd quarter of 2018.

COLLABORATORS

- Bay Area Clean Water Agencies
- AXYS Analytical
- BASMAA Agencies
- Dredged Material Management Office

MICROPLASTIC

Background

Microplastic, commonly defined as plastic particles smaller than 5 mm, comes in a broad range of polymer types, shapes, and sizes. Differences in size, shape, and chemistry can affect the way microplastic particles move through the environment, and may modify their potential for toxicity. Information on the chemistry and morphology of particles can help to identify sources and the means to mitigate the impact. The RMP has developed a Microplastic Science and Monitoring Strategy that articulates priority management questions and a multi-year plan for monitoring. The RMP is supporting a major microplastic monitoring and modeling effort funded primarily by the Gordon and Betty Moore Foundation. As part of this project, SFEI research partner 5 Gyres will develop data-driven regional recommendations for actions to reduce levels of microplastic in the Bay and adjacent national marine sanctuaries.

RELATION TO PERMIT REQUIREMENTS

There are no current permit requirements for microplastic, although large plastic items (> 5 mm) that may fragment into microplastic are addressed in the MRP and the statewide trash amendments and requirements.

USES OF RMP MICROPLASTIC DATA FOR DECISIONS

- Regional or state bans on single use plastic items and foam packaging materials

- State and Federal bans on microbeads

- Statewide trash amendments and requirements

- Public outreach and education regarding pollution prevention

PRIORITY QUESTIONS

1 How much microplastic pollution is there in the Bay?

2 What are the health risks?

3 What are the sources, pathways, loadings, and processes leading to microplastic pollution in the Bay?

4 Have the concentrations of microplastic in the Bay increased or decreased?

5 What management actions may be effective in reducing microplastic pollution?

Sampling with the manta trawl. Photograph by Shira Bezalel. ▶

MICROPLASTIC

RECENT FINDINGS

All field sampling for the Moore Foundation project has been successfully completed, with a total of 354 samples collected and sent to the University of Toronto for analysis. The analytical team has developed methods for extracting microplastic from a variety of matrices, including surface water, wastewater effluent, stormwater, sediment, and fish, and has refined methods for identifying microplastic polymer types using Raman and infrared spectroscopy.

Preliminary results suggest that fibers are ubiquitous across all matrices. Based on the data received to date, surface water Manta trawl samples frequently have high particle counts that consist largely of fibers. In samples that underwent spectroscopy, a majority of these fibers were identified as plastic.

As part of this project, we have developed a Bay transport model that links with oceanic transport models outside the Bay to evaluate transport of microplastic from the Bay to the Gulf of the Farallones. The microplastic monitoring data will be used to calibrate and validate this model.

WORKPLAN HIGHLIGHTS

Microplastic study results and policy recommendations: In 2019, SFEI and partners will complete all elements of the Moore Foundation microplastic project currently underway. This includes characterizing microplastic in Bay water, sediment, and prey fish; Bay Area stormwater and wastewater effluent; and open ocean water off the California coast. In addition, the project includes development of a transport model to predict the movement of microplastic from the Bay to adjacent national marine sanctuaries, and regional policy recommendations based on our results. SFEI will distribute a report, educational materials, and policy recommendations via a public symposium in 2019.

Microplastic monitoring in Bay bivalves: In 2018, the RMP will collect Bay bivalves for microplastic analyses. Results will be reported in 2019.

Update of San Francisco Bay Microplastic Science and Monitoring Strategy: Based on the wealth of new data, the RMP will update the strategy document and multi-year plan outlining priorities for future work.

COLLABORATORS

- Gordon and Betty Moore Foundation
- 5 Gyres Institute
- University of Toronto
- University of Michigan
- San Francisco Baykeeper
- Wylie Charters
- Moss Landing Marine Laboratories
- Patagonia
- Bay Area Clean Water Agencies
- East Bay Municipal Utility District
- City of Palo Alto
- Plus M Productions

TREATMENT EFFICIENCY

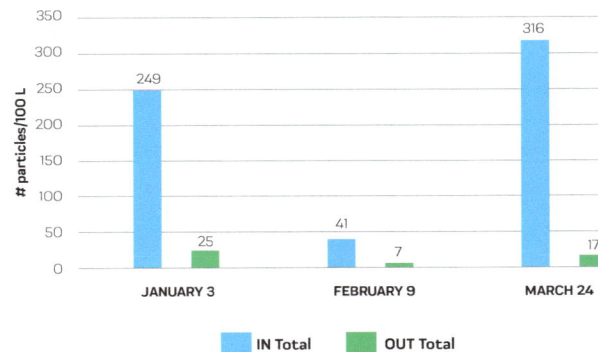

Bar chart. Y-axis: # particles/100 L (0 to 350). X-axis categories: JANUARY 3, FEBRUARY 9, MARCH 24. IN Total (blue): January 3 = 249, February 9 = 41, March 24 = 316. OUT Total (green): January 3 = 25, February 9 = 7, March 24 = 17.

SEDIMENT

NOTE: The mission of the Sediment Workgroup is to provide technical oversight and stakeholder guidance on RMP studies addressing questions about sediment delivery, sediment transport, dredging, and beneficial reuse of sediment.

Background

Sediment is a critical water quality parameter for the Bay. Sediment transport is a major factor in the fate and transport of priority pollutants such as PCBs and mercury. Suspended sediment concentrations in the water are also important for preventing large algae blooms despite high nutrient concentrations.

The RMP has been studying sediment since the Program began in 1993. In recent years, sea level rise has heightened the interest in sediment supply to the Bay. The mass balance of sediment in the Bay is a critical factor for marshes and other shoreline habitats to be able to withstand the rising seas. As the San Francisco Bay Restoration Authority decides how to spend $500 million for habitat restoration, it is critical to know how much sediment will be available and where it will be available.

In 2018, the RMP created a new Sediment Workgroup to bring together key stakeholders and scientists studying this issue and to prioritize science studies to inform management decisions.

RELATION TO PERMIT REQUIREMENTS

Essential Fisheries Habitat Consultation, PCBs TMDL, Mercury TMDL

- Provides information for setting dredged material testing thresholds and in-Bay disposal limits.

Long-Term Management Strategy for Dredged Material in SF Bay

- Provides information about sediment mass balance in the whole Bay, subembayments, and margin areas.

USES OF RMP SMALL TRIBUTARY LOADING DATA FOR DECISIONS

- NOAA 2011 Programmatic Essential Fish Habitat Agreement & 2015 LTMS Amended Programmatic Biological Opinion

- Long-Term Management Strategy for Dredged Material in SF Bay (LTMS) to comply with the Basin Plan

- Regional Restoration Plans

- PCBs TMDL

- Mercury TMDL

PRIORITY QUESTIONS

1 What are acceptable levels of chemicals in sediment for placement in the Bay, baylands, or restoration projects?

2 Are there effects on fish, benthic species, and submerged habitats from dredging or placement of sediment?

3 What are the sources, sinks, pathways, and loadings of sediment and sediment-bound contaminants to and within the Bay and subembayments?

4 How much sediment is passively reaching tidal marshes and restoration projects and how could the amounts be increased by management actions?

5 What are the concentrations of suspended sediment in the Estuary and its segments?

SEDIMENT

RECENT FINDINGS

In Water Years (WYs) 2016 and 2017, the USGS monitored the sediment flux through the Golden Gate. This flux is the largest unknown in the sediment budget for the Bay. Results indicate that sediment loads from the Delta during winter storms were mostly retained in San Pablo Bay, even during the historically high floods of WY2017. One recommendation from the report was to use modeling to evaluate cumulative fluxes over longer periods than can be monitored.

USGS monitoring of suspended sediments at the Dumbarton Bridge in WY2016 indicated that particle flocculation is an important factor for accurately calculating the sediment flux into Lower South Bay. The RMP has allocated funds for a special study in 2018-2019 to follow-up on this finding.

A synthesis report estimated that net average annual sediment supply to San Francisco Bay from terrestrial sources during the most recent 22-year period (WY1995-2016) was 1.95 billion kilograms. Approximately 63% of the sediment supply was estimated to be from small tributaries that drain directly to the Bay. Net supply from the Central Valley (measured at Mallard Island) was approximately 37% of the total supply. Bedload supply, after accounting for dredging, removals, and storage in flood control channels, was essentially zero. Recent data do not indicate any trends besides the step decrease in supply from the Delta in 1999. The report contains initial recommendations for improvements in sediment supply monitoring.

WORKPLAN HIGHLIGHTS

The RMP is funding the USGS to continue to measure suspended sediment flux at Mallard Island in the Delta. The sediment flux at this location is the largest single input of sediment to the Bay. USGS has measured sediment flux at this site since water year 1994. The RMP support is critical for maintaining a long-term record of sediment supply to the Bay until federal funding can be restored.

USGS has deployed new sensors to measure sediment fluxes at multiple depths and flocculation at the Dumbarton Bridge. The new equipment will be used to test hypotheses about flocculation processes and to more accurately measure the net fluxes of sediment into Lower South Bay. A final report will be available in early 2019.

SFEI is developing a Sediment Monitoring Strategy that the RMP can use to prioritize monitoring needs. The work will be completed as part of the EPA-funded Healthy Watersheds-Resilient Baylands Project, which is tackling similar issues. The RMP has provided matching funds to expand the scope of this strategy, which will be prepared in the fall of 2019.

Sediment in the Van Veen grab sampler. Photograph by Shira Bezalel. ▶

COLLABORATORS

- US Geological Survey
- US Army Corps of Engineers
- Bay Conservation and Development Commission
- San Francisco Bay Regional Water Quality Control Board
- Bay Planning Commission
- US Environmental Protection Agency

ACKNOWLEDGEMENTS

The RMP Steering Committee, 2018.

RMP Staff

Shira Bezalel

Nina Buzby

Ariella Chelsky

Jay Davis

Amy Franz

Alicia Gilbreath

Cristina Grosso

Jennifer Hunt

Erika King

Diana Lin

Lester McKee

John Ross

Meg Sedlak

David Senn

Ila Shimabuku

Rebecca Sutton

Jennifer Sun

Philip Trowbridge

Michael Weaver

Taylor Winchell

Adam Wong

Jing Wu

Don Yee

Editors

Jay Davis

Phil Trowbridge

Contributing Authors

Jay Davis, Meg Sedlak, Dave Senn, Rebecca Sutton, Phil Trowbridge, Jing Wu, Don Yee

Report Design

Ruth Askevold

The following reviewers greatly improved this document by providing comments on draft versions:

Nirmela Arsem

Nina Buzby

Peter Carroll

Beth Christian

Mary Lou Esparza

Tom Hall

Lester McKee

Kelly Moran

Ila Shimabuku

Chris Sommers

Phil Trowbridge

Luisa Valiela

Leah Walker

Additional Photo Credits

Shira Bezalel - pages v, vi-1, 10-11, and 36-37.

RMP

REGIONAL MONITORING
PROGRAM FOR WATER QUALITY
IN SAN FRANCISCO BAY

sfei.org/rmp

The RMP is administered by the San Francisco Estuary Institute

4911 Central Avenue, Richmond, CA 94804, p: 510-746-SFEI (7334), f: 510-746-7300, www.sfei.org

www.ingramcontent.com/pod-product-compliance
Lightning Source LLC
Chambersburg PA
CBHW041259210326

41598CB00010B/848